A *Handbook* of
Statistical
Analyses
using Stata

A Handbook of
Statistical
Analyses
using Stata

Sophia Rabe-Hesketh
Brian Everitt

CHAPMAN & HALL/CRC

Boca Raton   London   New York   Washington, D.C.

Library of Congress has Cataloged the Hard Covered Imprint Edition as Follows:

Rabe-Hesketh, Sophia.
    Handbook of statistical analysis using stata / Sophia Rabe
-Hesketh, Brian Everitt.
        p. cm.
    Includes bibliographical references and index.
    ISBN 0-8493-0387-7 (alk. paper)
    1. Stata. 2. Mathematical statistics--Data processing.
I. Everitt, Brian. II. Title.
QA276.4.R33 1998
519.5′.0285′5369—dc21   98-44800
                                                              CIP

Preface

Stata is an exciting statistical package that can be used for many standard and non-standard methods of data analysis. Stata is particularly useful for modeling complex data from longitudinal studies or surveys and is therefore ideal for analyzing results from clinical trials or epidemiological studies. The extensive graphic facilities of the software are also valuable to the modern data-analyst. In addition, Stata provides a powerful programming language that enables 'tailor-made' analyses to be applied relatively simply. As a result, many Stata users are developing (and making available to other users) new programs reflecting recent developments in statistics which are frequently incorporated into the Stata package.

This handbook follows the format of its two predecessors, *A Handbook of Statistical Analyses Using S-Plus* and *A Handbook of Statistical Analyses Using SAS*. Each chapter deals with the analysis appropriate for a particular set of data. A brief account of the statistical background is included in each chapter but the primary focus is on how to use Stata and how to interpret results. Our hope is that this approach will provide a useful complement to the excellent but very extensive Stata manuals. All the datasets can be accessed on the World Wide Web at *http:/www.iop.bpmf.ac.uk/home/depts/bc/books/stata*.

Sophia Rabe-Hesketh would like to acknowledge the usefulness of the Stata Netcourses in the writing of this book.

S. Rabe-Hesketh
B. S. Everitt
London, April 1998

To my parents, Birgit and Georg Rabe
Sophia Rabe-Hesketh

To my wife, Mary Elizabeth
Brian S. Everitt

.

Contents

Distributors for Stata

The distributors for Stata in the United Kingdom are

Timberlake Consultants
47 Hartfield Crescent
West Wickham
Kent BR4 9DW

email: info@timberlake.co.uk
web-site: http://www.timberlake.co.uk
telephone: 44 (0) 181 462 0495/0093

In the USA the distributors are

Stata Corporation
702 University Drive East
College Station, TX 77840

email: stata@stata.com
web-site: http://www.stata.com
telephone: 900 - 782 8272

CHAPTER 1

A Brief Introduction to Stata

1.1 Getting help and information

Stata is a general-purpose statistics package developed and maintained by Stata Corporation. There are several forms of Stata, "Intercooled Stata," its shorter version "Short Stata," and a simpler-to-use (point and click) student package "StataQuest." There are versions of each of these packages for Windows (3.1/3.11, 95, and NT), Unix platforms, and the Macintosh. In this book we will describe Intercooled Stata for Windows although most features are shared by the other versions of Stata.

The Stata package is described in five manuals (*Stata Getting Started, Stata User's Guide,* and *Stata Reference Manuals 1-3*) and in Hamilton (1998). The reference manuals provide extremely detailed information on each command while the User's guide describes Stata more generally. Features that are specific to the operating system are described in the appropriate *Getting Started* manual, e.g., *Getting Started with Stata for Windows.*

Each Stata-command has associated with it a help-file that may be viewed within a Stata session using the help facility. If the required command-name for a particular problem is not known, a list of possible command-names for that problem can be obtained using "lookup." Both the help-files and manuals refer to the Stata reference manuals by "[R] command name" and to the User's guide by "[U] chapter number."

The Stata Web-page (http://www.stata.com) contains information on the Stata mailing list and Internet courses and has links to files containing extensions and updates of the package (see Section 1.10) as well as a "frequently asked questions" (FAQ) list and further information on Stata.

The Stata mailing list, "Statalist," simultaneously functions as a technical support service with Stata staff frequently offering very helpful responses to questions. The Statalist messages are archived at *http://www.hsph.harvard.edu/statalist.*

Internet courses, called "netcourses," take place via a temporary mailing list for course organizers and "attenders." Each week, the course organizers send out lecture notes and exercises that the attenders can discuss with each other until the organizers send out the answers to the exercises and to the questions raised by attenders.

1.2 Running Stata

When Stata is started, a window opens as shown in Figure 1.1 containing four child-windows labeled

- Stata Command
- Stata Results
- Review
- Variables

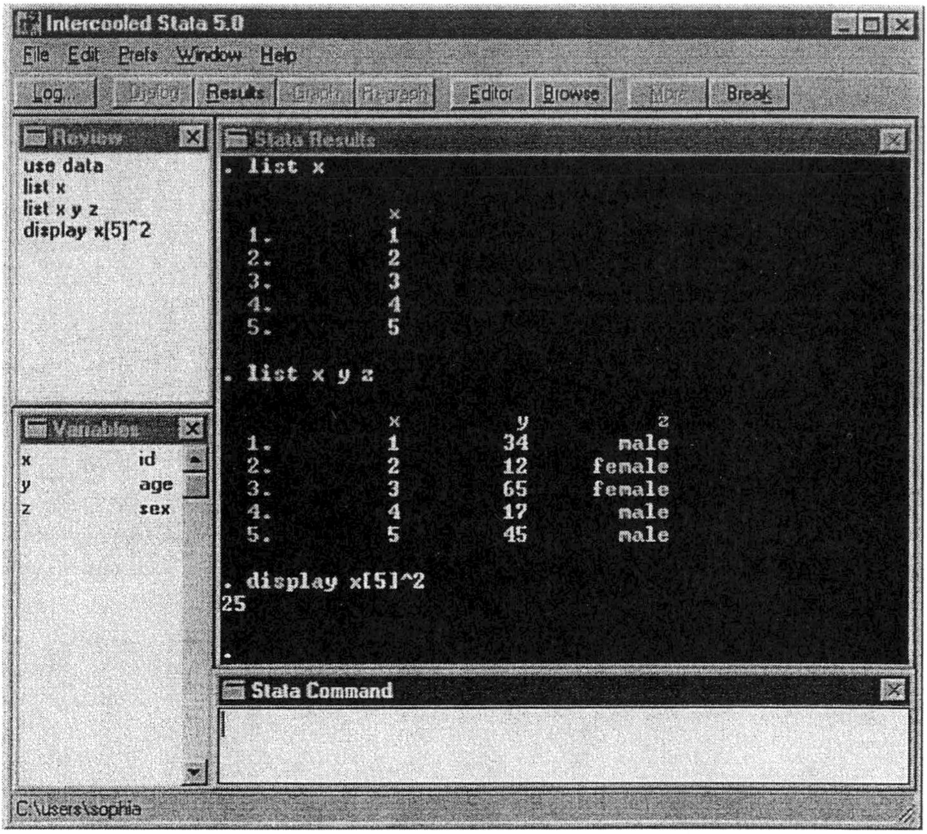

Figure 1.1 *Stata windows.*

When a period (the Stata prompt) appears in the "Stata Results" window, a command can be typed in the "Stata Command" window and executed by pressing the Return (or Enter) key. The command then appears next to the full stop in the "Stata Results" window, followed by the output. If the output is longer than the "Stata Results" window, --more-- appears at the bottom of the screen. Typing any key except b scrolls the output forward one page and typing b scrolls it back one page. However, b cannot be used to view output from previous commands.

Stata is ready to accept new commands when the prompt (i.e., a period) appears at the bottom of the screen. If Stata is not ready to receive new commands because it is still running or has not yet displayed all the current output, it may be interrupted by holding down *Ctrl* and pressing the Pause/Break key.

A previous command can be accessed using the "PgUp" and "PgDn" buttons or by selecting it from the "Review" window where all commands from the current Stata session are listed. The command can then be edited if required before pressing Return to execute the command.

A single dataset can be loaded into Stata (see Section 1.3). As in other statistical packages, this dataset is generally a matrix where the columns represent variables (with names and labels) and the rows represent observations. When a dataset is open, the variable names and variable labels appear in the "Variables" window.

Most Stata commands refer to a list of variables, the basic syntax being

```
command varlist
```

For example, if the dataset contains variables x y and z, then

```
list x y
```

lists the values of x and y. Other components can be added to the command, for example, adding if exp after varlist causes the command to process only those observations satisfying the logical expression exp. Options are separated from the main command by a comma. The complete command structure and its components are described in Section 1.4.

Although Stata is mainly based on commands, it also has a few "point and click" facilities. The main Stata window has the following entries in its top menu bar: **File, Edit, Prefs, Window,** and **Help.** Clicking into **Help** brings up the dialog box shown in Figure 1.2. Typing a topic, such as "survival," into the box,

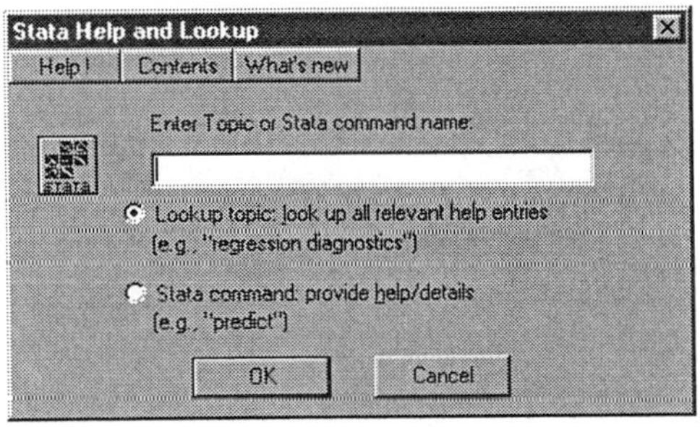

Figure 1.2 *Dialog box for help.*

selecting "Lookup topic," and pressing OK opens up a window containing a list of relevant command names or topics for which help files are available. Each entry in this list includes a green keyword (a hyperlink) that may be selected to view the appropriate help file. The same help file can be viewed directly by typing the keyword, for example "cox," into the help dialog box and selecting "Stata command." Each help file contains hyperlinks to other relevant help files. The lookup and help files can also be accessed using the commands

```
lookup survival
help cox
```

except that the files now appear in the "Stata Results" window where no hyperlinks are available.

Each of the Stata windows can be resized and moved around in the usual way. The fonts in a window can be changed by clicking into the menu button ▧ on the top left of that window's menu bar. All these settings can then be saved by selecting "Save Windowing Prefences" from the **Prefs** menu.

The tool bar underneath the menu bar of the main window has various options of which **Log, Editor,** and **Break** are the most important. The **Log** button enables a log file to be opened which can be viewed at any time during a Stata session to revisit previous output and accessed after the Stata session once it has been closed. The **Editor** button opens up a spreadsheet containing the loaded dataset (if any) enabling new data to be typed in or old data to be edited, and the **Break** button can be used (instead of *Ctrl* Break) to make Stata stop what it is doing.

Stata can be exited in three ways:

- Click into the close button ▣ at the top right hand corner of the Stata window,

- Select the **File** menu from the menu bar and select **Exit,**

- Type `exit, clear` in the "Stata Commands" window and press Return.

1.3 Datasets in Stata

Data input and output

Stata has its own data format with default extension *.dta*. Reading and saving a Stata file are straightforward. If the filename is *bank.dta*, the commands are

```
use bank
save bank
```

If the data are not stored in the current directory, then the complete path must be specified, as in the following command:

```
use c:\user\data\bank
```

However, the least error prone way of keeping all the files for a particular project in one directory is to change to that directory and save and read all files without their pathname:

```
cd c:\user\data
use bank
save bank
```

When reading a file into Stata, all data already in memory need to be cleared, either by running `clear` before the `use` command or by using the option `clear` as follows:

```
use bank,clear
```

If we wish to save data under an existing filename, this results in an error message unless we use the option `replace` as follows:

```
save bank, replace
```

If the data are not available in Stata format, they may be converted to Stata format using another package (e.g., Stat/Transfer) or saved as an ASCII file (although the latter options mean losing all the labels). When saving data as ASCII, missing values should be replaced by some numerical code.

There are three functions available for reading different types of ASCII data: `insheet`, the least flexible program, is for files containing one observation (on all variables) per line with variables separated by tabs or commas, where the first line may contain the variable names; `infile` with varlist (free format) allows line breaks to occur anywhere and variables to be separated by spaces as well as commas or tabs; `infile` with a dictionary (fixed format) is the most flexible program. Data can be saved as ASCII using `outfile` or `outsheet`.

Only one dataset can be loaded at any given time, but a dataset can be merged with the currently loaded dataset using the command `merge` to add observations or variables.

Variables

There are essentially two types of variables in Stata: string and numeric. Each variable can be one of a number of storage types that require different numbers of bytes. The storage types are byte, int, long, and float for numeric variables and str1 to str80 for string variables of different lengths. Besides the storage type, variables have associated with them a name, a label, and a format. The name of a variable y can be changed to x using

```
rename y x
```

The variable label can be defined using

```
label variable x "cost in pounds"
```

and the format of a numeric variable can be set to "general numeric" with two decimal places using

```
format x %7.2g
```

Numeric variables

Missing values in numeric variables are represented by dots only and are interpreted as very large numbers (which can lead to mistakes). Missing value codes can be converted to missing values using the command mvdecode. For example,

```
mvdecode x, mv(-99)
```

replaces all values of variable x equal to -99 by dots and

```
mvencode x, mv(-99)
```

changes the missing values back to -99.

Numeric variables can be used to represent categorical or continuous variables, including dates that are defined as the number of days since 1/1/1960. However, for categorical variables it is not always easy to remember which numerical code represents which category. Value labels can therefore be defined as follows:

```
label define s 1 married 2 divorced 3 widowed 4 single
value labels marital s
```

The categories can also be recoded, for example,

```
recode marital 2/3=2 4=3
```

merges categories 2 and 3 into category 2 and changes category 4 to 3.

Dates can be displayed using the date format %d. For example, listing the variable *time* in %7.0g format gives

```
list time
```

	time
1.	14976
2.	200

which is not as easy to interpret as

```
format time %d
list time
```

	time
1.	01jan01
2.	19jul60

String variables

String variables are typically used for categorical variables or in some cases for dates (e.g., if the file was saved as an ASCII file from SPSS). In Stata it is generally advisable to represent both categorical variables and dates by numeric variables, and conversion from string to numeric in both cases is straightforward. A categorical string variable can be converted to a numeric variable using the command `decode`, which replaces each unique string by an integer and uses that string as the label for the corresponding integer value. The command `encode` converts the labeled numeric variable back to a string variable.

A string variable representing dates can be converted to numeric using the function `date(string1, string2)` where `string1` is a string representing a date and `string2` is a permutation of "dmy" to specify the order of the day, month, and year in `string1`. For example, the commands

```
display date("30/1/1930","dmy")
```

and

```
display date("january 1, 1930", "mdy")
```

both return the negative value -10957 because the date is 10957 days before 1/1/1960.

1.4 Stata commands

Typing `help syntax` gives the following generic command structure for most Stata commands.

```
[by varlist:] command [varlist] [weight] [if exp] [in range] [using filename]
                      [, options]
```

The components have the following meaning:

[by varlist:] instructs Stata to repeat the command for each combination of values in the list of variables `varlist`.

[command] is the name of the command and can be abbreviated; for example, the command `display` can be abbreviated as `dis`.

[varlist] is the list of variables to which the command applies.

[weight] allows weights to be associated with observations (see Setion 1.6).

[if exp] restricts the command to that subset of the observations that satisfies the logical expression `exp`.

[in range] restricts the command to those observations whose indices lie in a particular `range`.

[using filename] specifies the filename to be used.

[options] are specific to the command and can be abbreviated.

For any given command, some of these components may not be available; for example, list does not allow [using filename]. The help-files for specific commands specify which components are available, using the same notation as above, with square brackets enclosing components that are optional. For example, help log gives

```
log using filename [, noproc append replace ]
```

implying that [by varlist:] is not allowed and that using filename is required, whereas the three options noproc, append, or replace are optional.

The syntax for varlist, exp, and range is described in the next three subsections, followed by information on how to loop through sets of variables or observations.

Varlist

The simplest form of varlist is a list of variable names separated by spaces. Variable names can also be abbreviated as long as this is unambiguous, i.e., $x1$ may be referred to by x only if there is no other variable starting on x such as x itself or $x2$. A range of consecutively numbered (or adjacent) variables such as $m1$, $m2$, and $m3$ may be referred to as m1-m3. All variables starting on the same set of letters can be represented by that set of letters followed by a wild card *, so that m* may stand for m1 m6 mother. The set of all variables is referred to by _all. Examples of a varlist are

```
x y
x1-x16
a1-a3 my* sex age
```

Expressions

There are logical, algebraic, and string expressions in Stata. Logical expressions evaluate to 1 (true) or 0 (false) and use the operators < and <= for "less than" and "less than or equal to," respectively, and similarly > and >= for "greater than" and "greater than or equal to." The symbols == and ~= stand for "equal to" and "not equal to," and the characters ~, &, and | represent "not," "and," and "or," respectively, so that

```
if (y~=2&z>x)|x==1
```

means "if y is not equal to two and z is greater than x or if x equals one." In fact, expressions involving variables are evaluated for each observation so that the expression really means

$$(y_i \neq 2 \& z_i > x_i)|x_i == 1$$

where i is the observation index.

Algebraic expressions use the usual operators + - * / and ^ for powers. Stata also has many mathematical functions such as sqrt() exp() log(), etc. and statistical functions such as chiprob() and normprob() for cumulative distribution functions and invnorm(), etc. for inverse cumulative distribution functions. Pseudo-random numbers can be generated using uniform(). An example of an algebraic expression is

```
invnorm(uniform())^2+3*exp(z)/y
```

where invnorm(uniform()) returns a (different) sample from the standard normal distribution for each observation.

Finally, string expressions mainly use special string functions such as substr(str,n1,n2) to extract a substring. The logical symbols == and ~= are also allowed with string variables and the operator + concatinates two strings. For example, the combined logical and string expression

```
("moon"+substr("sunlight",4,5))=="moonlight"
```

returns the value 1 for "true."

For a list of all functions, use help functions.

Observation indices and ranges

Each observation has associated with it an index. For example, the value of the third observation on a particular variable x may be referred to as x[3]. The macro _n takes on the value of the running index and _N is equal to the number of observations. We can therefore refer to the previous observation of a variable as x[_n-1].

An indexed variable is only allowed on the right hand side of an assignment. If we wish to replace x[3] by 2, we therefore need to use the syntax

```
replace x=2 if _n==3
```

We can refer to a range of observations using either if with a logical expression involving _n or, more easily by using in range, where range is a range of indices specified using the syntax f/l (for "first to last"), where f and/or l can be replaced by numerical values if required, so that 5/12 means "fifth to twelfth" and f/10 means "first to tenth," etc. Negative numbers are used to count from the end; for example,

```
list x in -10/1
```

lists the last 10 observations.

Looping through variables or observations

Explicitly looping through observations is often not necessary because expressions involving variables are automatically evaluated for each observation. It may however be required to repeat a command for subsets of observations and this is what by `varlist`: is for. Before using by `varlist`:, however, the data must be sorted using

```
sort varlist
```

where varlist includes the variables to be used for by `varlist`:. Note that if `varlist` contains more than one variable, ties in the earlier variables are sorted according to the next variable. For example,

```
sort school class
by school class: summ test
```

gives the summary statistics of *test* for each class. If *class* is labeled from 1 to n_i for the ith school, then not using `school` in the above commands would result in the observations for all classes labeled 1 to be grouped together.

A very useful feature of by `varlist`: is that it causes the observation index _n to go from 1 to _N within each group defined by `varlist`. For example,

```
sort group age
by group: list age if _n==_N
```

lists *age* for the last observation in each group where the last observation in this case is the observation with the highest age within its group.

We can also loop through a set variables or observations using `for`. For example,

```
for v*: list @
```

loops through the list of all variables starting on v and applies the command `list` to each element @ of the variable list. Numeric lists can also be used but need to be specified as such using `ltype(numeric)`;

```
for 1 3 5, ltype(numeric): list v@
```

lists *v1*, *v3*, and *v5*. Numeric lists can be abbreviated by "first-last/increment," giving the abbreviation 1-5/2 for the list 1 3 5. The `for` command can be made to loop through several lists (of the same length) simultaneously where the "current" elements of the different lists are referred to by @1, @2, etc. For example,

```
for v1-v5\1-5,ltype(varlist numeric): replace @1[@2]=0
```

replaces the ith value of the ith variable by 0, i.e., it sets vi[i] to 0. Since `for` loops simultaneously through all lists, it cannot be used to construct nested loops (see Section 1.9 for a description of `while`, which can be used for nested loops).

1.5 Data management

Generating variables

New variables can be generated using the commands `generate` or `egen`. The command `generate` simply equates a new variable to an expression that is evaluated for each observation. For example,

```
generate percent = 100*(old - new)/old if old>0
```

generates the variable *percent* that is equal to the percentage decrease from *old* to *new* for each observation where *old* is positive and equal to missing otherwise. The function `replace` works in the same way as `generate` except that it allows an existing variable to be changed. For example,

```
replace percent = 0 if old<=0
```

changes the missing values in *percent* to zeros. The two commands above could be replaced by the single command

```
generate percent=cond(old>0, 100*(old-new)/old, 0)
```

where `cond` evaluates to the second argument if the first argument is true and to the third argument otherwise.

The function `egen` provides an extension to `generate` with the following two advantages. The first advantage is that the expression on the right-hand side can be a function of a list of variables, whereas the functions for `generate` can only take simple expressions as arguments. For example, we can form the average of 100 variables *m1* to *m100* using

```
egen average=rmean(m1-m100)
```

where missing values are ignored. The second advantage is that the new variable can be a function of groups of observations. For example, if we have the income (variable *income*) for members within families (variable *family*), we may want to compute the total income of each member's family using

```
egen faminc = sum(income), by(family)
```

An existing variable can be replaced using egen functions only by first dropping it:

```
drop x
```

Another way of dropping variables is using `keep varlist`, where `varlist` is the list of all variables not to be dropped.

Changing the shape of the data

It is frequently necessary to change the shape of data, the most common application being grouped data, in particular repeated measures. If we have measurement occasion j for subject i, this can be viewed as a multivariate dataset in which each occasion j is represented by a variable xj and the subject identifier is in *subj*. However, for some statistical analysis, we may need one single, long, response vector containing all occasions for all subjects, as well as two variables *subj* and *occ* to represent the indices i and j, respectively. The two "data shapes" are called wide and long, respectively. We can convert from the wide shape with variables xj and *subj* given by

```
list
```

	x1	x2	subj
1.	2	3	1
2.	4	5	2

to the long shape with variables x, *occ* and *subj* using the syntax

```
reshape long x, i(subj) j(occ)
list
```

	subj	occ	x
1.	1	1	2
2.	1	2	3
3.	2	1	4
4.	2	2	5

and back again using

```
reshape wide x, i(subj) j(occ)
```

For data in the long shape, it may be necessary to compute summary measures from the repeated measures for each subject, for example, the mean, *meanx*, and standard deviation, *sdx* and the number of nonmissing repeated measures, *num*. This can be achieved using

```
collapse (mean) meanx=x (sd) sdx=x (count) num=x, by(subj)
```

Since it is not possible to convert back to the original format, the data may be preserved before running `collapse` and restored again later using the commands `preserve` and `restore`.

Other ways of changing the shape of data include dropping observations using

```
drop in 1/10
```

to drop the first 10 observations, or

```
sort group weight
by group: keep if _n==1
```

to drop all but the heaviest members of each group. Sometimes it is necessary to transpose the data, converting variables to observations and vice versa. This can be done and undone using `xpose`.

If each observation represents a number of units (as after collapse), it is sometimes necessary to replicate each observation by the number of units, *num*, that it represents. This may be done using

```
expand num
```

If there are two datasets, *subj.dta*, containing subject-specific variables, and *occ.dta*, containing occasion-specific variables for the same subjects, then if both files contain the same sorted subject identifier *subj_id* and *subj.dta* is currently loaded, the files can be merged as follows:

```
merge subj_id using occ
```

resulting in the variables from *group.dta* being expanded as in the expand command above and the variables from *occ.dta* being added.

1.6 Estimation

All estimation commands in Stata, for example, regress, logistic, poisson, and glm follow the same syntax and share many of the same options. The estimation commands also produce essentially the same output and save the same information that can be processed using the same set of "post-estimation" commands.

The basic command structure is

```
[xi:] command depvar [model] [weights], options
```

which may be combined with by varlist:, if exp and in range. The response variable is specified by depvar and the explanatory variables by model. If categorical explanatory variables and interactions are required, using xi: at the beginning of the command enables special notation for model to be used. For example,

```
xi: regress resp i.x*y z
```

fits a regression model with the main effects of x, y, and z and the interaction $x \times y$ where x is treated as categorical and y and z as continuous. Stata adds all the required dummy variables to the dataset (see help xi for further details).

The syntax for the

```
[weights]
```

option is

```
weighttype=varname
```

where weighttype depends on the purpose of weighting the data. If the data are in the form of a table where each observation represents a group containing a total of *freq* observations, using [fweight=freq] is equivalent to running the same estimation command on the expanded dataset where each observation has been replicated *freq* times. If the observations have different standard deviations, for example, because they represent averages of different numbers of observations, then `aweights` is used with weights proportional to the reciprocals of the standard deviations. Finally, `pweights` is used for inverse probability weighting in surveys where the weights are equal to the inverse probability that each observation was sampled. (Another type of weights, `iweight`, is available for some estimation commands mainly for use by programmers).

All the results of an estimation command are stored and can be processed using post-estimation commands. For example, `predict` can be used to compute predicted values or different types of residuals for the observations in the present dataset and the commands `test`, `testparm`, and `lrtest` can be used to carry out various test on the regression coefficients.

The saved results can also be accesssed directly using the names of the appropriate *global macro*. For example, the regression coefficients are stored in global macros called _b[varname] and their covariance matrix is stored as VCE. In order to display the regression coefficient of x, simply type

```
display _b[x]
```

1.7 Graphics

There is one command, `graph`, which can be used to plot a large number of different graphs in a "Stata Graph" window that appears when the first graph is plotted. The basic syntax is `graph varlist, options`, where `options` are used to specify the type of graph. For example,

```
graph x, box
```

gives a boxplot of x and

```
graph y x, twoway
```

gives a scatter-plot with y on the y-axis and x on the x-axis. (The option `twoway` is not needed here because it is the default.) More than the minimum number of variables may be given. For example,

```
graph x y, box
```

gives two boxplots within one set of axes and

```
graph y z x, twoway
```

gives a scatter-plot of y and z against x with different symbols for y and z. The option `by(group)` can be used to plot graphs separately for each group. With the option `box`, this results in several boxplots within one set of axes; and with the option `twoway`, this results in several scatter-plots in the same graphics window and using the same axis ranges.

If the variables have labels, then these are used as titles or axis labels as appropriate. The `graph` command can be extended to specify axis-labeling, or to specify which symbols should be used to represent the points and how (or whether) the points are to be connected, etc. For example, in the command

```
graph y z x, toway s(io) c(l.) xlabel ylabel t1("scatter plot")
```

the `symbol()` option `s(io)` causes the points in y to be invisible (i) and those in z to be represented by small circles (o). The `connect()` option `c(l.)` causes the points in y to be connected by straight lines and those in z to be unconnected. Finally, the `xlabel` amd `ylabel` options cause the x- and y-axes to be labeled using round values (without these options, only the minimum and maximum values are labeled) and the title option `t1("scatter plot")` causes a main title to be added at the top of the graph (`b1()`, `l1()`, `r(1)` would produce main titles on bottom, left, and right and `t2()`, `b2()`, `l2()`, `r2()` would produce secondary titles on each of the four sides).

The entire graph must be produced in a single command. This means, for example, that if different symbols are to be used for different groups on a scatter-graph, then each group must be represented by a separate variable having non-missing values only for observations belonging to that group. For example, the commands

```
gen y1=y if group==1
gen y2=y if group==2
graph y1 y2 x, s(dp)
```

produce a scatter-plot where y is represented by diamonds (d) in group 1 and by plus (p) in group 2. See `help graph` for a list of all plotting symbols, etc.

1.8 Stata as a calculator

Stata can be used as a simple calculator using the command `display` followed by an expression, e.g.,

```
display sqrt(5*(11-3^2))
```

```
3.1622777
```

There are also a number of statistical functions that can be used without reference to any variables. These commands end in i, where i stands for "immediate command." For example, we can calculate the sample size required for an independent samples t-test to achieve 80% power to detect a significant difference at the 1% level of significance (2-sided) if the means differ by one standard deviation using

```
sampsi 1 2, sd(1) power(.8) alpha(0.01)
```

```
Estimated sample size for two-sample comparison of means

Test Ho: m1 = m2, where m1 is the mean in population 1
                      and m2 is the mean in population 2
Assumptions:

           alpha =     0.0100   (two-sided)
           power =     0.8000
             m1 =         1
             m2 =         2
            sd1 =         1
            sd2 =         1
          n2/n1 =      1.00

Estimated required sample sizes:

             n1 =        24
             n2 =        24
```

Similarly, *ttesti* can be used to carry out a t-test if the means, standard deviations, and sample sizes are given.

Results can be saved in local macros using the syntax

```
local a=exp
```

and used again by enclosing the macro name in quotes ` ' . For example,

```
local a=5
display sqrt('a')
```

```
2.236068
```

Matrices can also be defined and matrix algebra carried out interactively. The following commands define a matrix a, display it, and give its trace and its eigenvalues;

```
matrix a=(1,2\2,4)
matrix list a
```

```
symmetric a[2,2]
         c1   c2
r1    1
r2    2    4
```

```
dis trace(a)
```

```
5
```

```
matrix symeigen x v = a
matrix list v
```

```
v[1,2]
        e1    e2
r1       5     0
```

1.9 Brief introduction to programming

So far, we have described commands as if they would be run interactively. However, in practice, it is always useful to be able to repeat the entire analysis using a single command. This is important, for example, when a data entry error is detected after most of the analysis has already been carried out! In Stata, a set of commands stored as an ASCII file, called, for example, *analysis.do*, can be executed using the command

```
do analysis
```

where the ".do" extension is not needed as it is the default extension for files containing commands, also called "do-files." We strongly recommend that readers create do-files for any work in Stata, for example, for the exercises of this book.

One way of generating a do-file is to carry out the analysis interactively and save the commands, for example, by selecting "Save Review Contents" from the menu of the "Review" window. Any text-editor or word-processor can then be used to edit the commands. The following is a useful template for a do-file:

```
/* comment describing what the file does */
version 5.0
capture log close
log using filename, replace
set more off

command 1
command 2
etc.

log close
exit
```

We will explain each line in turn.

1. The "brackets" /* and */ cause Stata to ignore everything between them. Another way of "commenting out" lines of text is to start the lines with a simple *.

2. The command version 5.0 causes Stata to interpret all commands as if

we were running Stata version 5.0 even if, in the future, we have actually installed a later version in which some of these commands do not work anymore.

3. The `capture` prefix causes the do-file to continue running even if the command results in an error. The `capture log close` command therefore closes the current log file if one is open or returns an error message. (Another useful prefix is `quietly`, which suppresses all output, except error messages.)

4. The command `log using filename, replace` opens a log file, replacing the previous log file if it exists.

5. The command `set more off` causes all the output to scroll past automatically instead of waiting for the user to scroll through it manually.

6. After the analysis is complete, the log file is closed using `log close`.

7. The last statement, `exit`, is not necessary at the end of a file but can be used to cause Stata to stop runnning the do-file wherever it is placed.

As mentioned, variables, global macros (see Section 1.6), local macros, and matrices (see Section 1.8) can be used for storing and referring to constants and these are made use of extensively in programs. For example, we may wish to subtract the mean of x from x. Interactively, we would use

```
summarize x
```

to find out what the mean value is and then subtract that value from x. However, we should not type the value of the mean into a "do-file" because the result would no longer be valid if the data change. Instead, we can refer to the mean computed by `summarize` using the name of the global macro in which it is stored, in this case `_result(3)`, as follows;

```
quietly summarize x
gen xnew=x-_result(3)
```

Stata also uses macro names S_#, S_E_# and others which are accessed by prefixing the name with a $ (e.g., $S_#). In order to find out under what names constants are stored, see the "stored results" section for the command of interest in the reference manual. Another way of finding out which macro contains which result is to run

```
disp_res
```

which displays the current values of all `_result(#)` macros or

```
macro list
```

which lists the other macros and their contents.

If a local macro is defined without using the = sign, anything can appear on the right-hand side and typing the local macro name in single quotes has the same effect as typing whatever appeared on the right-hand side in the definition of the macro. For example, if we have a variable y, we can use the commands

```
local a y
disp " `a'[1] = " `a'[1]
```

$$\boxed{\texttt{y[1] = 4.6169958}}$$

Sometimes it is useful to define a program that can be called with arguments in much the same way as Stata's own commands. One advantage of programs is that they allow the use of while. For example, this is how the first three observations of variables x, y, and z might be displayed. First define the program "mylist"

```
program define mylist
while "`1'"~=""{   /* outer loop: loop through variables */
     local x `1'
     local i=1
     display "`x'"
     while `i'<=3{   /* inner loop: loop through observations */
          display `x'[`i']
          local i=`i'+1   /* next observation */
     }
     mac shift   /* next variable */
     display " "
}
end
```

and then run the program using the command

```
mylist x y z
```

The inner loop simply displays the '`i`'th element of the variable '`x`' for '`i`' from 1 to 3. The outer loop uses the macro '`1`' as follows; At the beginning, the macros '`1`', '`2`', and '`3`' contain the arguments x, y, and z, respectively, with which mylist was called. The command

```
mac shift
```

shifts the contents of '`2`' into '`1`' and those of '`3`' into '`2`', etc. Therefore, the outer loop steps through variables x, y, and z.

A program can be defined by typing it into the "Commands" window. This is almost never done in practice, however, a more useful method being to define the program within a do-file where it can easily be edited. Note that once the program has been loaded into memory (by running the program define commands), it has to be cleared from memory using program drop before it can be redefined. It is therefore useful to have the command

```
capture program drop mylist
```

in the do-file before the **program define** command, where **capture** ensures that the do-file continues running even if **mylist** is not yet loaded.

A program may also be saved in a separate file (containing only the program definition) of the same name as the program itself and having the extension .*ado*. The "ado-file" ("automatic do-file") can be executed simply by typing the name of the file. There is no need to load the program first. In fact, many of Stata's own commands are actually "ado-files" stored in the directory *c:\stata\ado*.

1.10 Keeping Stata up to date

Stata Corporation releases official updates of Stata every 2 months as part of the Stata Technical Bulletin (STB), which also contains programs from Stata users. The Stata updates can be obtained by downloading the latest STB file from one of the sites accessible from Stata's Web page. When this file is unzipped into a temporary directory, many new directories are created, each corresponding to a particular set of programs. One of the directories is called *stata* and this contains the official updates produced by Stata Corporation. Copy the contents of the *stata* directory into *c:\stata\ado*. This overwrites some existing files and you may wish to save the contents of *c:\stata\ado* in another temporary directory in case anything goes wrong. The command help **whatsnew** in later Stata sessions lists all the changes since the release of the present version of Stata.

The most recent STB contains all official updates since the release of the current version of Stata so that the official updates in previous STBs are obsolete. The only reason why older STBs may be of interest is because they may contain useful user-contributed functions. Information on such functions can be found using lookup. For example, running

```
lookup survival
```

gives a long list of entries including one on STB-35

```
STB-35   crc45 . . . . . . . . . . . . . .  New options for survival-time data
         (help stset)
         1/97    STB Reprints Vol 6, pages 7--8
         adds two new options to stset
```

which reveals that STB-35 has a directory in it called crc45 containing files for "New options for survival time data" and that help on these may be found using **help stset**. Not all user-defined programs are included in any STB (yet). Other ado-files may be found on the Stata Web site under "cool ados" or on the Statalist archive under "contributed ADO files" (direct address *http://ideas.uqam.ca/ideas/data/bocbocode.html*).

1.11 Exercises

1. Use an editor (e.g., Notepad, PFE or a word-processor) to generate the dataset *test.dat* given below, where the columns are separated by tabs (make sure to save it as a text only, or ASCII, file).

v1	v2	v3
1	3	5
2	16	3
5	12	2

2. Read the data into Stata using `insheet` (see `help insheet`).

3. Click into **Editor** and type in the variable *sex* with values 1 2 and 1.

4. Define value labels for sex (1=male, 2=female).

5. Use `gen` to generate *id*, a subject index (from 1 to 3).

6. Use `rename` to rename the variables *v1* to *v3* to *time1* to *time3*. Also try doing this in a single command using `for`.

7. Use `reshape` to convert the dataset to long shape.

8. Generate a variable *d* that is equal to the squared difference between the variable *time* at each occasion and the average of *time* for each subject.

9. Drop the observation corresponding to the third occasion for *id*=2.

CHAPTER 2

Data Description and Simple Inference: Female Psychiatric Patients

2.1 Description of data

The data to be used in this chapter consist of observations on eight variables for 118 female psychiatric patients and are available in Hand et al. (1994). The variables are as follows.

- *age*: age in years
- *IQ*: intelligence questionnaire score
- *anxiety*: anxiety (1=none, 2=mild, 3=moderate, 4=severe)
- *depress*: depression (1=none, 2=mild, 3=moderate, 4=severe)
- *sleep*: can you sleep normally? (1=yes, 2=no)
- *sex*: have you lost interest in sex? (1=no, 2=yes)
- *life*: have you thought recently about ending your life? (1=no, 2=yes)
- *weight*: weight change over last 6 months (in lb)

The data are given in Table 2.1 with missing values coded as −99. One question of interest is how the women who have recently thought about ending their lives differ from those who have not. Also of interest are the correlations between anxiety and depression and between weight change, age, and IQ.

Table 2.1 Data in *fem.dat*

id	age	IQ	anx	depress	sleep	sex	life	weight
1	39	94	2	2	2	2	2	4.9
2	41	89	2	2	2	2	2	2.2
3	42	83	3	3	3	2	2	4
4	30	99	2	2	2	2	2	−2.6
5	35	94	2	1	1	2	1	−0.3
6	44	90	−99	1	2	1	1	0.9
7	31	94	2	2	−99	2	2	−1.5
8	39	87	3	2	2	2	1	3.5
9	35	−99	3	2	2	2	2	−1.2
10	33	92	2	2	2	2	2	0.8
11	38	92	2	1	1	1	1	−1.9

Table 2.1 Data in *fem.dat*

12	31	94	2	2	2	−99	1	5.5
13	40	91	3	2	2	2	1	2.7
14	44	86	2	2	2	2	2	4.4
15	43	90	3	2	2	2	2	3.2
16	32	−99	1	1	1	2	1	−1.5
17	32	91	1	2	2	−99	1	−1.9
18	43	82	4	3	2	2	2	8.3
19	46	86	3	2	2	2	2	3.6
20	30	88	2	2	2	2	1	1.4
21	34	97	3	3	−99	2	2	−99
22	37	96	3	2	2	2	1	−99
23	35	95	2	1	2	2	1	−1
24	45	87	2	2	2	2	2	6.5
25	35	103	2	2	2	2	1	−2.1
26	31	−99	2	2	2	2	1	−0.4
27	32	91	2	2	2	2	1	−1.9
28	44	87	2	2	2	2	2	3.7
29	40	91	3	3	2	2	2	4.5
30	42	89	3	3	2	2	2	4.2
31	36	92	3	−99	2	2	2	−99
32	42	84	3	3	2	2	2	1.7
33	46	94	2	−99	2	2	2	4.8
34	41	92	2	1	2	2	1	1.7
35	30	96	−99	2	2	2	2	−3
36	39	96	2	2	2	1	1	0.8
37	40	86	2	3	2	2	2	1.5
38	42	92	3	2	2	2	1	1.3
39	35	102	2	2	2	2	2	3
40	31	82	2	2	2	2	1	1
41	33	92	3	3	2	2	2	1.5
42	43	90	−99	−99	2	2	2	3.4
43	37	92	2	1	1	1	1	−99
44	32	88	4	2	2	2	1	−99
45	34	98	2	2	2	2	−99	0.6
46	34	93	3	2	2	2	2	0.6
47	42	90	2	1	1	2	1	3.3
48	41	91	2	1	1	1	1	4.8
49	31	−99	3	1	2	2	1	−2.2
50	32	92	3	2	2	2	2	1
51	29	92	2	2	2	1	2	−1.2
52	41	91	2	2	2	2	2	4

Table 2.1 Data in *fem.dat*

53	39	91	2	2	2	2	2	5.9
54	41	86	2	1	1	2	1	0.2
55	34	95	2	1	1	2	1	3.5
56	39	91	1	1	2	1	1	2.9
57	35	96	3	2	2	1	1	−0.6
58	31	100	2	2	2	2	2	−0.6
59	32	99	4	3	2	2	2	−2.5
60	41	89	2	1	2	1	1	3.2
61	41	89	3	2	2	2	2	2.1
62	44	98	3	2	2	2	2	3.8
63	35	98	2	2	2	2	1	−2.4
64	41	103	2	2	2	2	2	−0.8
65	41	91	3	1	2	2	1	5.8
66	42	91	4	3	−99	−99	2	2.5
67	33	94	2	2	2	2	1	−1.8
68	41	91	2	1	2	2	1	4.3
69	43	85	2	2	2	1	1	−99
70	37	92	1	1	2	2	1	1
71	36	96	3	3	2	2	2	3.5
72	44	90	2	−99	2	2	2	3.3
73	42	87	2	2	2	1	2	−0.7
74	31	95	2	3	2	2	2	−1.6
75	29	95	3	3	2	2	2	−0.2
76	32	87	1	1	2	2	1	−3.7
77	35	95	2	2	2	2	2	3.8
78	42	88	1	1	1	2	1	−1
79	32	94	2	2	2	2	1	4.7
80	39	−99	3	2	2	2	2	−4.9
81	34	−99	3	−99	2	2	1	−99
82	34	87	3	3	2	2	1	2.2
83	42	92	1	1	2	1	1	5
84	43	86	2	3	2	2	2	0.4
85	31	93	−99	2	2	2	2	−4.2
86	31	92	2	2	2	2	1	−1.1
87	36	106	2	2	2	1	2	−1
88	37	93	2	2	2	2	2	4.2
89	43	95	2	2	2	2	1	2.4
90	32	95	3	2	2	2	2	4.9
91	32	92	−99	−99	−99	2	2	3
92	32	98	2	2	2	2	2	−0.3
93	43	92	2	2	2	2	2	1.2

Table 2.1 Data in *fem.dat*

94	41	88	2	2	2	2	1	2.6
95	43	85	1	1	2	2	1	1.9
96	39	92	2	2	2	2	1	3.5
97	41	84	2	2	2	2	2	−0.6
98	41	92	2	1	2	2	1	1.4
99	32	91	2	2	2	2	2	5.7
100	44	86	3	2	2	2	2	4.6
101	42	92	3	2	2	2	1	−99
102	39	89	2	2	2	2	1	2
103	45	−99	2	2	2	2	2	0.6
104	39	96	3	−99	2	2	2	−99
105	31	97	2	−99	−99	−99	2	2.8
106	34	92	3	2	2	2	2	−2.1
107	41	92	2	2	2	2	2	−2.5
108	33	98	3	2	2	2	2	2.5
109	34	91	2	1	1	2	1	5.7
100	42	91	3	3	2	2	2	2.4
111	40	89	3	1	1	1	1	1.5
112	35	94	3	3	2	2	2	1.7
113	41	90	3	2	2	2	2	2.5
114	32	96	2	1	1	2	1	−99
115	39	87	2	2	2	1	2	−99
116	41	86	3	2	1	1	2	−1
117	33	89	1	1	1	1	1	6.5
118	42	−99	3	2	2	2	2	4.9

2.2 Group comparison and correlations

We have interval scale variables (weight change, age and IQ), ordinal variables (anxiety and depression), and categorical, dichotomous variables (sex and sleep) that we wish to compare between two groups of women, those who have thought about ending their lives and those who have not.

For interval scale variables, the most common statistical test is the t-test, which assumes that the observations in the two groups are independent and are sampled from two populations each having a normal distribution and equal variances. A nonparametric alternative (which does not rely on the latter two assumptions) is the Mann-Whitney U-test.

For ordinal variables, either the Mann-Whitney U-test or a χ^2-test may be appropriate, depending on the number of levels of the ordinal variable. The latter test can also be used to compare dichotomous variables between the groups.

Continuous variables can be correlated using the Pearson correlation. If we are interested in the question whether the correlations differ significantly from zero, then a hypothesis test is available that assumes bivariate normality. A significance test not making this distributional assumption is available; it is based on the correlation of the ranked variables, the Spearman rank correlation. Finally, if variables have only few categories, Kendall's tau-b provides a useful measure of correlation (see, e.g., Sprent (1993)).

2.3 Analysis using Stata

Assuming the data have been saved from a spreadsheet or statistical package (e.g., SPSS) as a tab-delimited ASCII file, *fem.dat*, they can be read using the instruction

```
insheet using fem.dat, clear
```

There are missing values which have been coded as -99. We remove these using

```
mvdecode _all, mv(-99)
```

In order to have consistent coding for "yes" and "no," we recode the variable sleep

```
recode sleep 1=2 2=1
```

and, to avoid confusion in the future, we label the values as follows:

```
label define yn 1 no 2 yes
label values sex yn
label values life yn
label values sleep yn
```

The last three commands could also have been carried out in a `for` loop

```
for sex life sleep: label values @ yn
```

First, we can compare the groups who have and have not thought about ending their lives by tabulating summary statistics of various variables for the two groups. For example, for *IQ*, we type

```
table life, contents( mean iq sd iq)
```

```
----------+----------------------
  Life |    mean(iq)      sd(iq)
----------+----------------------
    no |    91.27084     3.757204
   yes |    92.09836       5.0223
----------+----------------------
```

In order to assess whether the groups appear to differ in their weight loss over the past 6 months and to informally check assumptions for an independent samples t-test, we plot the variable *weight* as a boxplot for each group after defining appropriate labels:

```
label variable weight "weight change in last six months"
label variable life /*
    */ "have you recently thought about ending your life?"
graph weight, box by(life)
```

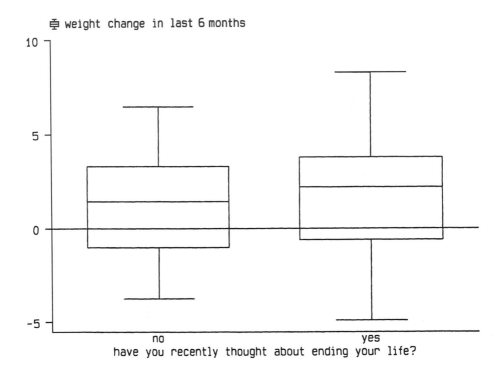

Figure 2.1 *Boxplot of weight by group.*

giving the graph shown in Figure 2.1. (Note that in the instructions above, the "brackets" for comments, /* and */ were used to make Stata ignore the line breaks in the middle of the *label* command.) The groups do not seem to differ much in their mean weight change and the assumptions for the t-test seem reasonable because the distributions are symmetric with similar spread.

We can also check the assumption of normality more formally by plotting a normal quantile plot of suitably defined residuals. Here the difference between the observed weight changes and the group-specific mean weight changes can be used. If the normality assumption is satisfied, the quantiles of the residuals should be linearly related to the quantiles of the normal distribution. The residuals can be computed and plotted using

```
egen res=mean(weight), by(life)
replace res=weight-res
label variable res "residuals of t-test for weight"
qnorm res, gap(5) xlab ylab t1("normal q-q plot")
```

where gap(5) was used to reduce the gap between the vertical axis and the axis title (the default gap is 8). The points in the Q-Q plot in Figure 2.2 are sufficiently close to the staight line.

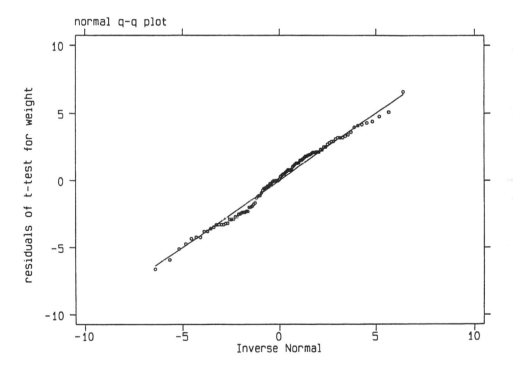

Figure 2.2 *Normal Q-Q plot of residuals of weight change.*

We could also test whether the variances differ significantly using

```
sdtest weight, by(life)
```

```
Variance ratio test

  ----------------------------------------------------------------------
  Group |    Obs      Mean    Std. Err.   Std. Dev.   [95% Conf. Interval]
  --------+-------------------------------------------------------------
     no |     45   1.408889   .3889616    2.609234   .6249883     2.19279
    yes |     61   1.731148   .3617847    2.825629   1.00747     2.454825
  --------+-------------------------------------------------------------
combined |    106   1.59434    .2649478    2.727805   1.068997    2.119682
  ----------------------------------------------------------------------

                       Ho: sd(no) = sd(yes)

           F(44,60) observed   = F_obs           =    0.853
           F(44,60) lower tail = F_L   = F_obs   =    0.853
           F(44,60) upper tail = F_U   = 1/F_obs =    1.173

  Ha: sd(no) < sd(yes)        Ha: sd(no) ~= sd(yes)      Ha: sd(no) > sd(yes)
    P < F_obs = 0.2919     P < F_L + P > F_U = 0.5724     P > F_obs = 0.7081
```

showing that there is no significant difference ($p = 0.57$) between the variances. Note that the test for equal variances is only appropriate if the variable can be assumed to be normally distributed in each population. Having checked the assumptions, we carry out a t-test for weight change:

```
ttest weight, by(life)
```

```
Two-sample t test with equal variances

  ----------------------------------------------------------------------
  Group |    Obs      Mean    Std. Err.   Std. Dev.   [95% Conf. Interval]
  --------+-------------------------------------------------------------
     no |     45   1.408889   .3889616    2.609234   .6249883     2.19279
    yes |     61   1.731148   .3617847    2.825629   1.00747     2.454825
  --------+-------------------------------------------------------------
combined |    106   1.59434    .2649478    2.727805   1.068997    2.119682
  --------+-------------------------------------------------------------
   diff |          -.3222587   .5376805               -1.388499    .743982
  ----------------------------------------------------------------------
Degrees of freedom: 104

               Ho: mean(no) - mean(yes) = diff = 0

    Ha: diff < 0              Ha: diff ~= 0              Ha: diff > 0
      t =  -0.5993              t =  -0.5993              t =  -0.5993
    P < t =   0.2751       P > |t| =   0.5502          P > t =   0.7249
```

The two-tailed significance is $p = 0.55$ and the 95% confidence interval for the mean difference in weight change between those who have thought about ending their lives and those who have not is from -0.74 pounds to 1.39 pounds. Therefore, there is no evidence that the populations differ in their weight change, but we also cannot rule out a mean difference in weight change of as much as 1.4 pounds in 6 months.

We use a χ^2-test to test for differences in depression between the two groups and display the corresponding cross-tabulation together with the percentage of women in each category of depression by group using the single command

```
tab life depress, row chi2
```

```
          | Depress
     Life |         1          2          3 |     Total
----------+---------------------------------+----------
       no |        26         24          1 |        51
          |     50.98      47.06       1.96 |    100.00
----------+---------------------------------+----------
      yes |         0         42         16 |        58
          |      0.00      72.41      27.59 |    100.00
----------+---------------------------------+----------
    Total |        26         66         17 |       109
          |     23.85      60.55      15.60 |    100.00

          Pearson chi2(2) =   43.8758    Pr = 0.000
```

There is a significant association between *depress* and *life* with none of the subjects who have thought about ending their lives having zero depression compared with 51% of those who have not. Note that this test does not take account of the ordinal nature of depression and is therefore likely to be less sensitive than, for example, ordinal regression (see Chapter 6). Fisher's exact test can be obtained without having to reproduce the table as follows:

```
tab life depress, exact nofreq
```

```
          Fisher's exact =                0.000
```

Similarly, for sex, we can obtain the table and both the χ^2-test and Fisher's exact test using

```
tab life sex, row chi2 exact
```

```
          | Sex
     Life |        no        yes |     Total
----------+----------------------+----------
       no |        12         38 |        50
          |     24.00      76.00 |    100.00
----------+----------------------+----------
      yes |         5         58 |        63
          |      7.94      92.06 |    100.00
----------+----------------------+----------
    Total |        17         96 |       113
          |     15.04      84.96 |    100.00

                 Pearson chi2(1) =    5.6279    Pr = 0.018
                 Fisher's exact =                0.032
          1-sided Fisher's exact =                0.017
```

Therefore, those who have thought about ending their lives are more likely to have lost interest in sex than those who have not (92% compared with 76%). We can explore correlations between the three variables *weight*, *IQ*, and *age* using a single command

```
corr weight iq age
```

```
(obs=100)

        |   weight        iq       age
--------+---------------------------------
 weight |   1.0000
     iq |  -0.2920    1.0000
    age |   0.4131   -0.4363    1.0000
```

The correlation matrix has been evaluated for those 100 observations that had complete data on all three variables. The command `pwcorr` can be used to include, for each correlation, all observations that have complete data for the corresponding pair of variables resulting in different sample sizes for different correlations. The p-value and sample sizes can be displayed simultaneously as follows:

```
pwcorr weight iq age, obs sig
```

```
          |   weight        iq       age
----------+---------------------------------
   weight |   1.0000
          |
          |     107
          |
       iq |  -0.2920    1.0000
          |   0.0032
          |     100       110
          |
      age |   0.4156   -0.4345    1.0000
          |   0.0000    0.0000
          |     107       110       118
          |
```

The corresponding scatter-plot matrix is obtained using

```
graph weight iq age, matrix half jitter(1)
```

where `jitter(1)` randomly moves the points by a very small amount to stop them overlapping completely due to the discrete nature of age and IQ. The resulting graph is shown in Figure 2.3. Thus, older psychiatric female patients tend to put on more weight than younger ones, as do less intelligent women. However, older women in this sample also tended to be less intelligent so that age and intelligence are confounded.

It would be interesting to see whether those who have thought about ending their lives have the same relationship between age and weight change as those

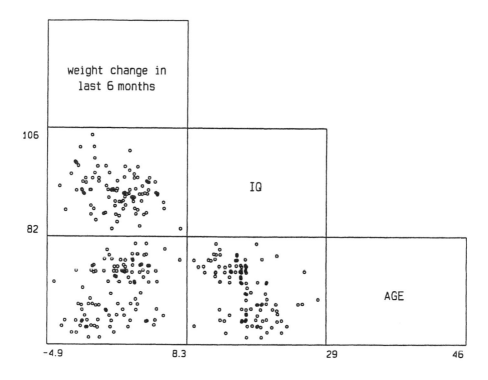

Figure 2.3 *Scatter-plot matrix for weight, IQ, and age.*

who have not. In order to form a single scatter-plot with different symbols representing the two groups, we must use a single variable for the *x*-axis (*age*) and plot two separate variables *wgt1* and *wgt2* that contain the weight changes for groups 1 and 2, respectively:

```
gen wgt1 = weight if life==2
gen wgt2 = weight if life==1
label variable wgt1 "no"
label variable wgt2 "yes"
graph wgt1 wgt2 age, s(dp) xlabel ylabel        /*
     */ ll("weight change in last 6 months")  /*
     */ saving(wasct, replace)
```

The resulting graph in Figure 2.4 shows that within both groups, higher age is associated with larger weight increases and the groups do not form distinct clusters.

Finally, an appopriate correlation between depression and anxiety is Kendall's tau-b, which can be obtained using

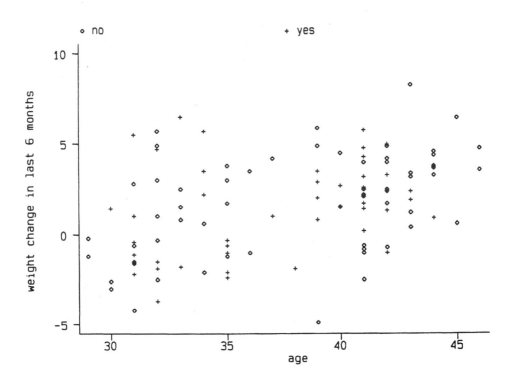

Figure 2.4 *Scatter-plot of weight against age.*

```
ktau depress anxiety
```

```
    Number of obs =      107
Kendall's tau-a =        0.2827
Kendall's tau-b =        0.4951
Kendall's score =     1603
     SE of score =      288.279   (corrected for ties)

Test of Ho: depress and anxiety independent
       Pr > |z| =        0.0000   (continuity corrected)
```

giving a value of 0.50 with an approximate p-value of $p < 0.001$.

2.4 Exercises

1. Tabulate the mean weight change by level of depression.

2. Using `for`, tabulate the means and standard deviations by *life* for each of the variables *age*, *iq*, and *weight*.

3. Use `lookup nonparametric` or `lookup mann` or `lookup whitney` to find help on how to run the Mann-Whitney U-test.

4. Compare the weight changes between the two groups using the Mann Whitney U-test.

5. Form a scatter-plot for *IQ* and *age* using different symbols for the two groups (life=1 and life=2). Explore the use of the option `jitter(#)` for different integers # to stop symbols overlapping.

6. Having tried out all these commands interactively, create a do-file containing these commands and run the do-file. Use the option `saving(filename,replace)` to save the graphs and view the graphs later using the command `graph using filename`.

See also exercises in Chapter 6.

CHAPTER 3

Multiple Regression: Determinants of Pollution in U.S. Cities

3.1 Description of data

The data to be analyzed in this chapter were collected by Sokal and Rohlf (1981) from several U.S. government publications, and are also available in Hand et al. (1994). The data relate to air pollution in 41 U.S. cities; they are reproduced here in Table 3.1. There is a single dependent variable, *so2*, the annual mean concentration of sulphur dioxide, in micrograms per cubic meter. These data generally relate to means for the 3 years 1969–1971. The values of six explanatory variables, two of which concern human ecology and four climate, are also recorded; details are as follows:

- *temp*: average annual temperature in °F
- *manuf*: number of manufacturing enterprises employing 20 or more workers
- *pop*: population size (1970 census) in thousands
- *wind*: average annual wind speed in miles per hour
- *precip*: average annual precipitation in inches
- *days*: average number of days with precipitation per year

The main question of interest about this data is how the pollution level as measured by sulphur dioxide concentration is determined by the six explanatory variables. The central method of analysis will be *multiple regression*.

3.2 The multiple regression model

The multiple regression model has the general form

$$y_i = \beta_0 + \beta_1 x_{1i} + \beta_2 x_{2i} + \cdots + \beta_p x_{pi} + \epsilon_i \qquad (3.1)$$

where y is a continuous response variable, $x_1, x_2, \cdots, x_p$ are a set of explanatory variables and ϵ is a residual term. The regression coefficients, $\beta_0, \beta_1, \cdots, \beta_p$ are generally estimated by least squares. Significance tests for the regression coefficients can be derived by assuming that the residual terms are normally distributed with zero mean and constant variance σ^2.

For n observations of the response and explanatory variables, the regression

Table 3.1 Data in *usair.dat*

Town	SO$_2$	temp	manuf	pop	wind	precip	days
Phoenix	10	70.3	213	582	6.0	7.05	36
Lrock	13	61.0	91	132	8.2	48.52	100
Sfran	12	56.7	453	716	8.7	20.66	67
Denver	17	51.9	454	515	9.0	12.95	86
Hartford	56	49.1	412	158	9.0	43.37	127
Wilming	36	54.0	80	80	9.0	40.25	114
Washing	29	57.3	434	757	9.3	38.89	111
Jackson	14	68.4	136	529	8.8	54.47	116
Miami	10	75.5	207	335	9.0	59.80	128
Atlanta	24	61.5	368	497	9.1	48.34	115
Chicago	110	50.6	3344	3369	10.4	34.44	122
Indian	28	52.3	361	746	9.7	38.74	121
DesM	17	49.0	104	201	11.2	30.85	103
Wichita	8	56.6	125	277	12.7	30.58	82
Louisv	30	55.6	291	593	8.3	43.11	123
NewO	9	68.3	204	361	8.4	56.77	113
Baltim	47	55.0	625	905	9.6	41.31	111
Detroit	35	49.9	1064	1513	10.1	30.96	129
Minn	29	43.5	699	744	10.6	25.94	137
Kansas	14	54.5	381	507	10.0	37.00	99
StLouis	56	55.9	775	622	9.5	35.89	105
Omaha	14	51.5	181	347	10.9	30.18	98
Alburq	11	56.8	46	244	8.9	7.77	58
Albany	46	47.6	44	116	8.8	33.36	135
Buffalo	11	47.1	391	463	12.4	36.11	166
Cincinn	23	54.0	462	453	7.1	39.04	132
Cleve	65	49.7	1007	751	10.9	34.99	155
Colum	26	51.5	266	540	8.6	37.01	134
Philad	69	54.6	1692	1950	9.6	39.93	115
Pittsb	61	50.4	347	520	9.4	36.22	147
Provid	94	50.0	343	179	10.6	42.75	125
Memphis	10	61.6	337	624	9.2	49.10	105
Nashville	18	59.4	275	448	7.9	46.00	119
Dallas	9	66.2	641	844	10.9	35.94	78
Houston	10	68.9	721	1233	10.8	48.19	103
SLC	28	51.0	137	176	8.7	15.17	89
Norfolk	31	59.3	96	308	10.6	44.68	116
Richmond	26	57.8	197	299	7.6	42.59	115
Seattle	29	51.1	379	531	9.4	38.79	164
Charlest	31	55.2	35	71	6.5	40.75	148
Milwak	16	45.7	569	717	11.8	29.07	123

model can be written concisely as

$$\mathbf{y} = \mathbf{X}\boldsymbol{\beta} + \boldsymbol{\epsilon} \qquad (3.2)$$

where $\mathbf{y}$ is the $n \times 1$ vector of responses, $\mathbf{X}$ is an $n \times (p+1)$ matrix of known constants, the first column containing a series of ones corresponding to the term β_0 in (3.1) and the remaining columns values of the explanatory variables. The elements of the vector $\boldsymbol{\beta}$ are the regression coefficients $\beta_0, \cdots, \beta_p$, and those of the vector $\boldsymbol{\epsilon}$, the residual terms $\epsilon_1, \cdots, \epsilon_n$. For full details of multiple regression see, for example, Rawlings (1988).

3.3 Analysis using Stata

Assuming the data are available as an ASCII file *usair.dat* in the current directory and that the file contains city names as given in Table 3.1, they can be read in for analysis using the following instruction:

```
infile str 10 town so2 temp manuf pop /*
      */ wind precip days using usair.dat
```

Before undertaking a formal regression analysis of these data, it will be helpful to examine them graphically using a scatter-plot matrix. Such a display is useful in assessing the general relationships between the variables, in identifying possible outliers, and in highlighting potential collinearity problems among the explanatory variables. The basic plot can be obtained using

```
graph so2 temp manuf pop wind precip days, matrix
```

The resulting diagram is shown in Figure 3.1. Several of the scatter-plots show evidence of outliers and the relationship between *manuf* and *pop* is very strong, suggesting that using both as explanatory variables in a regression analysis may lead to problems (see later). The relationships of particular interest, namely those between *so2* and the explanatory variables (the relevant scatter-plots are those in the first row of Figure 3.1) indicate some possible nonlinearity.

A more informative, although slightly more 'messy' diagram can be obtained if the plotted points are labeled with the associated town name. The necessary Stata instruction is

```
graph so2-days, matrix symbol([town]) tr(3) ps(150)
```

The symbol option labels the points with the names in the *town* variable; if, however, the full name is used, the diagram would be very difficult to read. Consequently, the `trim` option is used to select the first three characters of each name for plotting and the `psize` option is used to increase the size of these characters to 150% compared with the usual 100% size. The resulting diagram appears in Figure 3.2. Clearly, Chicago and to a lesser extent Philadelphia might be considered outliers. Chicago has such a high degree of pollution compared to the other cities that it should perhaps be considered as a special case and

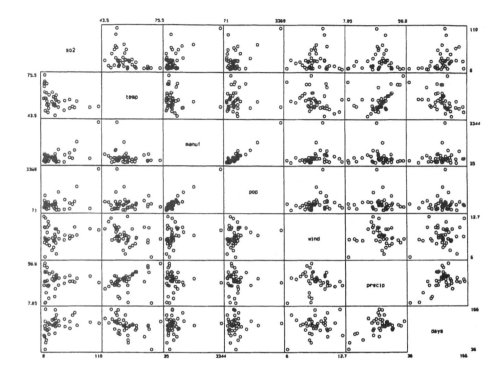

Figure 3.1 *Scatter-plot matrix.*

excluded from further analysis. A new data file with Chicago removed can be
generated using

```
clear
infile str 10 town so2 temp manuf pop wind /*
    */ precip days using usair.dat if town~="Chicago"
```

or by dropping the opservation using

```
drop if town=="Chicago"
```

Stata has a number of procedures that can be used to fit the basic multiple regression model; for example, fit and regress. Here, the former will be
used since then many useful regression diagnostics can be easily accessed. The
necessary Stata instruction to regress sulphur dioxide concentration on the six
explanatory variables is

```
fit so2 temp manuf pop wind precip days
```

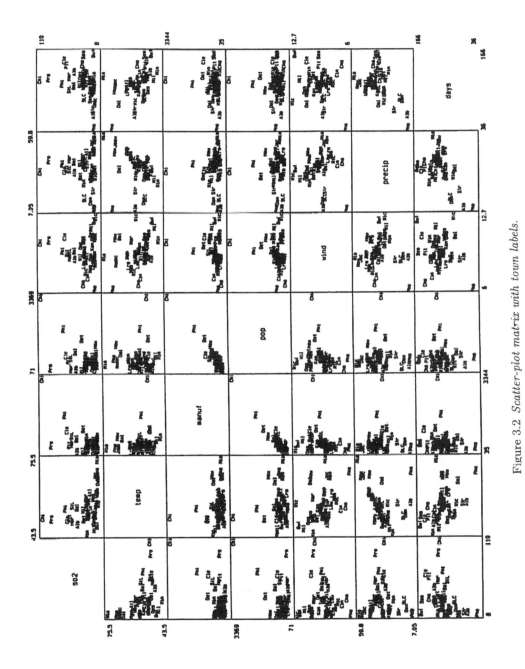

Figure 3.2 Scatter-plot matrix with town labels.

or, alternatively,

```
fit so2 temp-days
```

```
  Source |       SS        df       MS              Number of obs =       40
---------+------------------------------            F(  6,    33) =      6.20
   Model | 8203.60523       6  1367.26754           Prob > F      =   0.0002
Residual | 7282.29477      33  220.675599           R-squared     =   0.5297
---------+------------------------------            Adj R-squared =   0.4442
   Total |  15485.90       39  397.074359           Root MSE      =   14.855
------------------------------------------------------------------------------
     so2 |    Coef.   Std. Err.       t     P>|t|      [95% Conf. Interval]
---------+--------------------------------------------------------------------
    temp | -1.268452   .6305259    -2.012   0.052    -2.551266      .0143631
   manuf |  .0654927   .0181777     3.603   0.001     .0285098      .1024756
     pop | -.039431    .0155342    -2.538   0.016    -.0710357     -.0078264
    wind | -3.198267   1.859713    -1.720   0.095    -6.981881      .5853469
  precip |  .5136846   .3687273     1.393   0.173    -.2364966     1.263866
    days | -.0532051   .1653576     -.322   0.750    -.3896277      .2832175
   _cons |  111.8709   48.07439     2.327   0.026     14.06278      209.679
------------------------------------------------------------------------------
```

The main features of interest are the analysis of variance table and the parameter estimates. In the former, the ratio of the model mean square to the residual mean square gives an F-test for the hypothesis that *all* the regression coefficients in the fitted model are zero. The resulting F-statistic with 6 and 33 degrees of freedom takes the value 6.20 and is shown on the right-hand side; the associated p-value is very small. Consequently the hypothesis is rejected. The square of the multiple correlation coefficient (R^2) is 0.52 showing that 52% of the variance of sulphur dioxide concentration is accounted for by the six explanatory variables of interest. The adjusted R^2 statistic is an estimate of the population R^2, taking account of the fact that the parameters were estimated from the data. The statistic is calculated as

$$\text{adj } R^2 = 1 - \frac{(n-i)(1-R^2)}{n-p} \tag{3.3}$$

where n is the number of observations used in fitting the model, and i is an indicator variable that takes the value 1 if the model includes an intercept and 0 otherwise. The root MSE is simply the square root of the residual mean square in the analysis of variance table, which itself is an estimate of the parameter σ^2. The estimated regression coefficients give the estimated change in the response variable produced by a unit change in the corresponding explanatory variable with the remaining explanatory variables held constant.

One concern generated by the initial graphical material on this data was the strong relationship between the two explanatory variables *manuf* and *pop*. The correlation of these two variables is obtained using

```
corr manuf pop
```

```
(obs=40)
         |    manuf       pop
---------+--------------------
  manuf|    1.0000
    pop|    0.8906    1.0000
```

The strong linear dependence might be a source of collinearity problems and can be investigated further by calculating what are known as *variance inflation factors* for each of the explanatory variables. These are given by

$$\text{VIF}(x_i) = \frac{1}{1 - R_i^2} \tag{3.4}$$

where $\text{VIF}(x_i)$ is the variance inflation factor for explanatory variable x_i and R_i^2 is the square of the multiple correlation coefficient obtained from regressing x_i on the remaining explanatory variables.

The variance inflation factors can be found in Stata by following `fit` with `vif`:

```
vif
```

```
Variable |      VIF       1/VIF
---------+----------------------
  manuf |     6.28     0.159275
    pop |     6.13     0.163165
   temp |     3.72     0.269156
   days |     3.47     0.287862
 precip |     3.41     0.293125
   wind |     1.26     0.790619
---------+----------------------
Mean VIF |     4.05
```

Chatterjee and Price (1991) give the following 'rules-of-thumb' for evaluating these factors:

• Values larger than 10 give evidence of multicollinearity.

• A mean of the factors considerably larger than one suggests multicollinearity.

Here, there are no values greater than 10 (as an exercise we suggest readers also calculate the VIFs when the observations for Chicago are included), but the mean value of 4.05 gives some cause for concern. A simple (although not necessarily the best) way to proceed is to drop one of *manuf* or *pop*. We will exclude the former and repeat the regression analysis using the five remaining explanatory variables:

```
fit so2 temp pop wind precip days
```

```
  Source |       SS         df       MS              Number of obs =      40
---------+------------------------------             F(  5,    34) =    3.58
   Model | 5339.03465        5  1067.80693           Prob > F      =  0.0105
Residual | 10146.8654       34  298.437216           R-squared     =  0.3448
---------+------------------------------             Adj R-squared =  0.2484
   Total | 15485.90         39  397.074359           Root MSE      =  17.275
----------------------------------------------------------------------------
     so2 |    Coef.    Std. Err.      t     P>|t|     [95% Conf. Interval]
---------+------------------------------------------------------------------
    temp | -1.867665   .7072827    -2.641   0.012    -3.305037    -.430294
     pop |  .0113969   .0075627     1.507   0.141    -.0039723    .0267661
    wind | -3.126429   2.16257     -1.446   0.157    -7.5213      1.268443
  precip |  .6021108   .4278489     1.407   0.168    -.2673827    1.471604
    days | -.020149    .1920012     -.105   0.917    -.4103424    .3700445
   _cons |  135.8565   55.36797     2.454   0.019     23.33529    248.3778
----------------------------------------------------------------------------
```

```
vif
```

```
Variable |     VIF      1/VIF
---------+--------------------
    days |    3.46    0.288750
    temp |    3.46    0.289282
  precip |    3.40    0.294429
    wind |    1.26    0.790710
     pop |    1.07    0.931015
---------+--------------------
Mean VIF |    2.53
```

The variance inflation factors are now satisfactory.

The very general hypothesis concerning all regression coefficients mentioned above is not usually of great interest in most applications of multiple regression since it is most unlikely that all the chosen explanatory variables will be unrelated to the response variable. The more interesting question is whether a subset of the regression coefficients is zero, implying that not all the explanatory variables are of use in determining the response variable. A preliminary assessment of the likely importance of each explanatory variable can be made using the table of estimated regression coefficients and associated statistics. Using a conventional 5% criterion, the only 'significant' coefficient is that for the variable *temp*. Unfortunately, this very simple approach is not in general suitable, since in most cases the explanatory variables are correlated. Consequently, removing a particular variable from the regression will alter both the estimated regression coefficients of the remaining variables and their standard errors. A more involved approach to identifying important subsets of explanatory variables is therefore required. A number of procedures are available.

1. Considering some *a priori* ordering or grouping of the variables generated by the substantive questions of interest.

2. Automatic selection methods, which are of the following types:

   a. *Forward selection* This method starts with a model containing some of the explanatory variables and then considers variables one by one for inclusion. At each step, the variable added is the one that results in the biggest increase in the model or regression sum of squares. An F-type statistic is used to judge when further additions would not represent a significant improvement in the model.

   b. *Backward elimination* Here, variables are considered for removal from an initial model containing all the explanatory variables. At each stage, the variable chosen for exclusion is the one leading to the smallest reduction in the regression sum of squares. Again, an F-type statistic is used to judge when further exclusions would represent a significant deterioration in the model.

   c. *Stepwise regression* This method is essentially a combination of the previous two. The forward selection procedure is used to add variables to an existing model and, after each addition, a backward elimination step is introduced to assess whether variables entered earlier might now be removed, because they no longer contribute significantly to the model.

In the best of all possible worlds, the final model selected by the three automatic procedures would be the same. This is often the case, but it is not guaranteed. It should also be stressed that none of the automatic procedures for selecting subsets of variables are foolproof. They must be used with care, and warnings such as those given in McKay and Campbell (1982a, 1982b) concerning the validity of the F-tests used to judge whether variables should be included or eliminated, noted.

In this example, begin by considering an *a priori* grouping of the five explanatory variables since one, *pop*, relates to human ecology and the remaining four to climate. To perform a forward selection procedure with the ecology variable treated as a single term (all variables being either entered or not entered based on their joint significance) and similarly the climate terms, requires the following instruction:

```
sw regress so2 (pop) (temp wind precip days), pe(0.05)
```

```
                     begin with empty model
p = 0.0119 <  0.0500  adding    temp wind precip days
  Source |      SS        df       MS                    Number of obs =       40
---------+------------------------------                 F(  4,    35) =     3.77
   Model | 4661.27545      4  1165.31886                 Prob > F      =   0.0119
Residual | 10824.6246     35  309.274987                 R-squared     =   0.3010
---------+------------------------------                 Adj R-squared =   0.2211
   Total |   15485.90     39  397.074359                 Root MSE      =   17.586

----------------------------------------------------------------------------------
    so2 |     Coef.   Std. Err.       t    P>|t|     [95% Conf. Interval]
--------+-------------------------------------------------------------------------
   temp | -1.689848    .7099204    -2.380   0.023    -3.131063   -.2486329
   wind | -2.309449    2.13119     -1.084   0.286    -6.635996    2.017097
 precip |  .5241595    .4323535     1.212   0.234    -.3535647    1.401884
   days |  .0119373    .1942509     0.061   0.951    -.382413     .4062876
  _cons |  123.5942    55.75236     2.217   0.033     10.41091    236.7775
----------------------------------------------------------------------------------
```

Note the grouping as required. The pe term specifies the significance level of the F-test for addition to the model. Terms with a p-value less than the figure specified will be included. Here, only the climate variables are shown since they are jointly significant ($p = 0.0119$) at the significance level for inclusion.

As an illustration of the automatic selection procedures, the following Stata instruction applies the backward elimination method, with explanatory variables whose F-values for removal have associated p-values greater than 0.2 being removed:

```
sw fit so2 temp pop wind precip days, pr(0.2)
```

```
                     begin with full model
p = 0.9170 >= 0.2000  removing days

  Source |      SS        df       MS                    Number of obs =       40
---------+------------------------------                 F(  4,    35) =     4.60
   Model | 5335.74801      4   1333.937                  Prob > F      =   0.0043
Residual | 10150.152      35  290.004343                 R-squared     =   0.3446
---------+------------------------------                 Adj R-squared =   0.2696
   Total |   15485.90     39  397.074359                 Root MSE      =    17.03


----------------------------------------------------------------------------------
    so2 |     Coef.   Std. Err.       t    P>|t|     [95% Conf. Interval]
--------+-------------------------------------------------------------------------
   temp | -1.810123    .4404001    -4.110   0.000    -2.704183   -.9160635
    pop |  .0113089    .0074091     1.526   0.136    -.0037323    .0263501
   wind | -3.085284    2.096471    -1.472   0.150    -7.341347    1.170778
 precip |  .5660172    .2508601     2.256   0.030     .0567441    1.07529
  _cons |  131.3386    34.32034     3.827   0.001     61.66458    201.0126
----------------------------------------------------------------------------------
```

With the chosen significance level, only the variable *days* is excluded.

The next stage in the analysis should be an examination of the *residuals* from the chosen model; that is, the differences between the observed and fitted values of sulphur dioxide concentration. Such a procedure is vital for assessing model assumptions, identifying any unusual features in the data, indicating outliers, and suggesting possibly simplifying transformations. The most useful ways of examining the residuals are graphical and the most commonly used plots are as follows.

- A plot of the residuals against each explanatory variable in the model. The presence of a curvilinear relationship, for example, would suggest that a higher-order term, perhaps a quadratic in the explanatory variable, should be added to the model.

- A plot of the residuals against predicted values of the response variable. If the variance of the response appears to increase with predicted value, a transformation of the response may be in order.

- A normal probability plot of the residuals—after all systematic variation has been removed from the data, the residuals should look like a sample from the normal distribution. A plot of the ordered residuals against the expected order statistics from a normal distribution provides a graphical check on this assumption.

The first two plots can be obtained after using the `fit` procedure with the `rvpplot` and `rvfplot` instructions. For example, for the model chosen by the backward selection procedure, a plot of residuals against predicted values with the first three letters of the town name used to label the points is obtained using the instruction

```
rvfplot, symbol([town]) tr(3) ps(150) xlab ylab gap(3)
```

The resulting plot is shown in Figure 3.3, and indicates a possible problem, namely the apparently increasing variance of the residuals as the fitted values increase (see also Chapter 7). Perhaps some thought needs to be given to the possible transformations of the response variable (see exercises).

Next, graphs of the residuals plotted against each of the four explanatory variables can be obtained using the following series of instructions

```
rvpplot pop, symbol([town]) tr(3) ps(150) xlab ylab
ll("Residuals") gap(3)
rvpplot temp, symbol([town]) tr(3) ps(150) xlab ylab
ll("Residuals") gap(3)
rvpplot wind, symbol([town]) tr(3) ps(150) xlab ylab
ll("Residuals") gap(3)
rvpplot precip, symbol([town]) tr(3) ps(150) xlab ylab
ll("Residuals") gap(3)
```

The resulting graphs are shown in Figures 3.4 to 3.7. In each graph, the point corresponding to the town *Providence* is somewhat distant from the bulk of the points, and the graph for *wind* has perhaps a 'hint' of a curvilinear structure.

The simple residuals plotted by `rvfplot` and `rvpplot` have a distribution that is scale dependent since the variance of each is a function of both σ^2 and

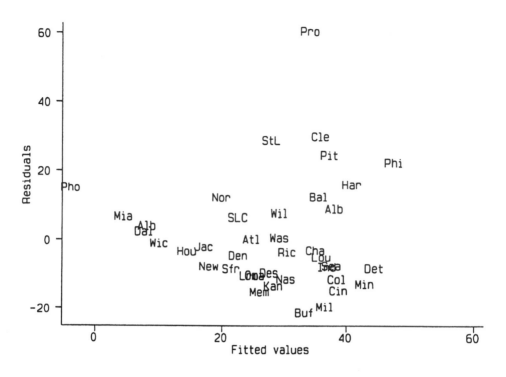

Figure 3.3 *Residuals against predicted response.*

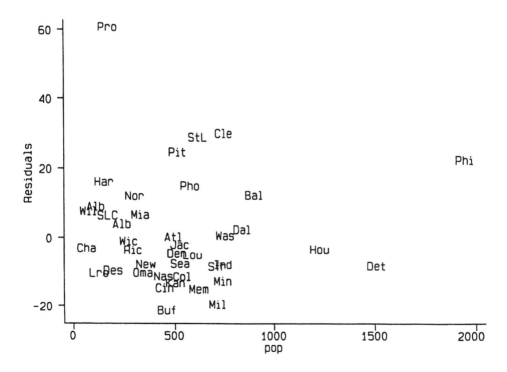

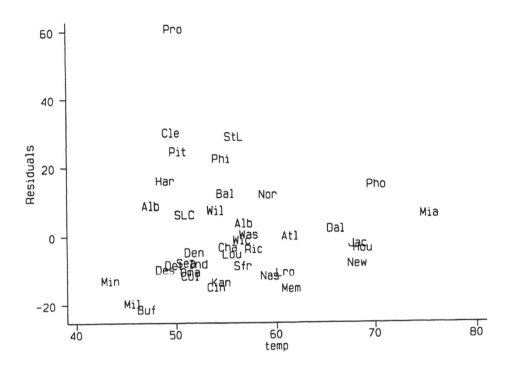

Figure 3.5 *Residuals against temperature.*

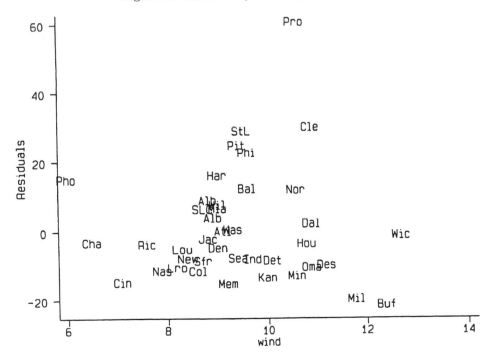

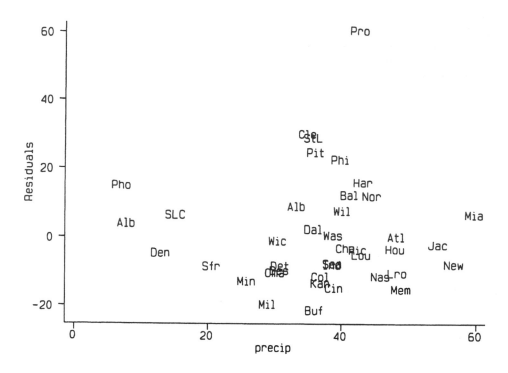

Figure 3.7 *Residuals against precipitation.*

the diagonal values of the so-called 'hat' matrix, $\mathbf{H}$, given by:

$$\mathbf{H} = \mathbf{X}(\mathbf{X'X})^{-1}\mathbf{X'}$$

(3.5)

(see Cook and Weisberg (1982) for a full explanation of the hat matrix). Consequently, it is often more useful to work with a standardized version, r_i calculated as follows:

$$r_i = \frac{y_i - \hat{y}_i}{s\sqrt{1 - h_{ii}}}$$

(3.6)

where s^2 is the estimate of σ^2, $\hat{y}_i$ is the predicted value of the response, and h_{ii} is the ith diagonal element of $\mathbf{H}$.

These standardized residuals can be obtained in the `fit` procedure by applying the `fpredict` instruction. For example, to obtain a normal probability plot of the standardized residuals and to plot them against the fitted values requires the following instructions:

```
fpredict fit
fpredict sdres, rstandard
pnorm sdres, gap(5)
graph sdres fit, symbol([town]) tr(3) ps(150) xlab ylab gap(3)
```

The first instruction stores the fitted values in the variable *fit*, the second stores the standardized residuals in the variable *sdres*, the third produces a normal probability plot (Figure 3.8), and the last instruction produces the graph of standardized residuals against fitted values, which is shown in Figure 3.9.

The normal probability plot indicates that the distribution of the residuals departs somewhat from normality. The pattern in the plot shown in Figure 3.9 is identical to that in Figure 3.3 but here values outside (-2,2) indicate possible outliers; in this case, the point corresponding to the town *Providence*. Analogous plots to those in Figures 3.4 to 3.7 could be obtained in the same way.

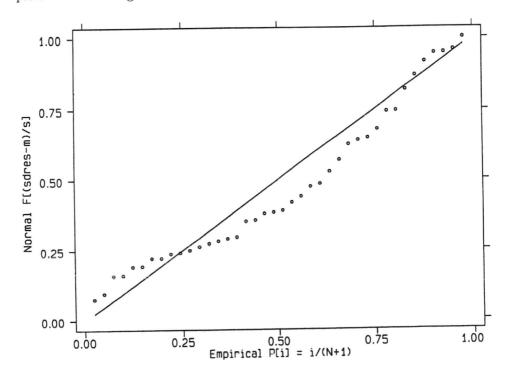

Figure 3.8 *Normal probability plot of standardized residuals.*

A rich variety of other diagnostics for investigating fitted regression models have been developed during the last decade and many of these are available with the fit procedure. Illustrated here is the use of two of these, namely the *partial residual plot* (Mallow, 1986) and *Cook's distance statistic* (Cook, 1977, 1979). The former is useful in identifying whether, for example, quadratic or other higher order terms are needed for any of the explanatory variables; the latter measures the change to the estimates of the regression coefficients that results from deleting each observation and can be used to indicate those observations that may be having an undue influence on the estimation and fitting process.

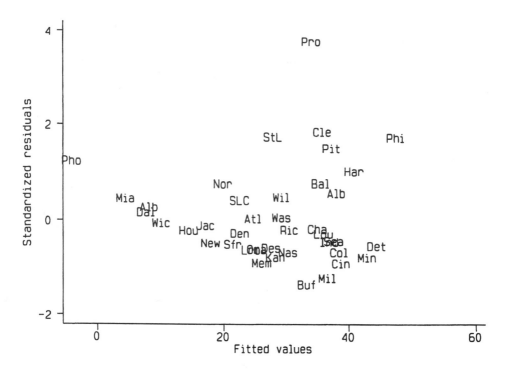

Figure 3.9 *Standardized residuals against predicted values.*

The partial residual plots are obtained in the `fit` procedure from the `cprplot` instruction. For the four explanatory variables in the selected model for the pollution data, the required plots are obtained as follows:

```
cprplot pop, border c(k) xlab ylab l1("Partial Residuals") gap(4)
cprplot temp, border c(k) xlab ylab l1("Partial Residuals") gap(4)
cprplot wind, border c(k) xlab ylab l1("Partial Residuals") gap(4)
cprplot precip, border c(k) xlab ylab l1("Partial Residuals") gap(4)
```

The two `graph` options used, `border` and `c(k)`, produce a border for the graphs and a locally weighted regression curve or *lowess*. The resulting graphs are shown in Figures 3.10 to 3.13. The graphs have to be examined for nonlinearities and whether the regression line, which has slope equal to the estimated effect of the corresponding explanatory variable in the chosen model, fits the data adequately. The added lowess curve is generally helpful for both. None of the four graphs give any obvious indication of nonlinearity.

The Cook's distance statistics are found by again using the `fpredict` instruction; the following calculates these statistics for the chosen model for the pollution data and lists the observations where the statistic is greater than 4/40 (4/n), which is usually the value regarded as indicating possible problems.

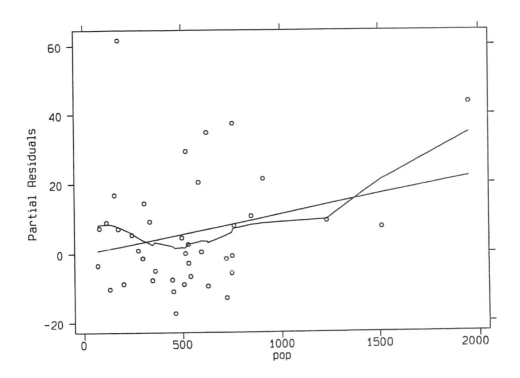

Figure 3.10 *Partial residual plot for population.*

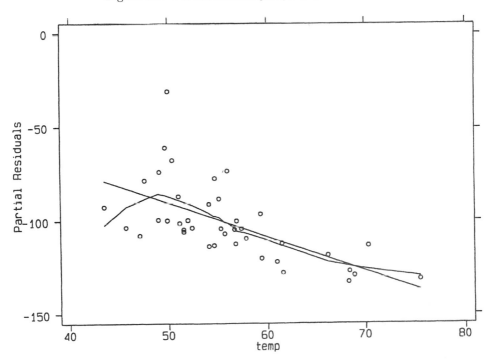

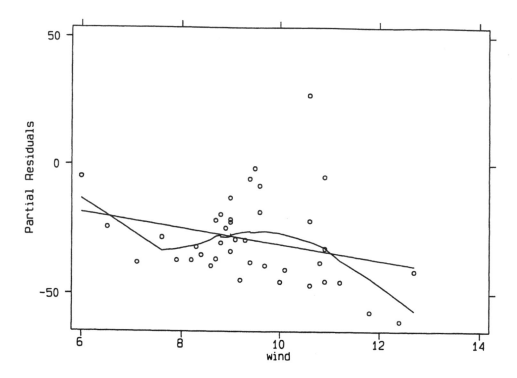

Figure 3.12 *Partial residual plot for wind speed.*

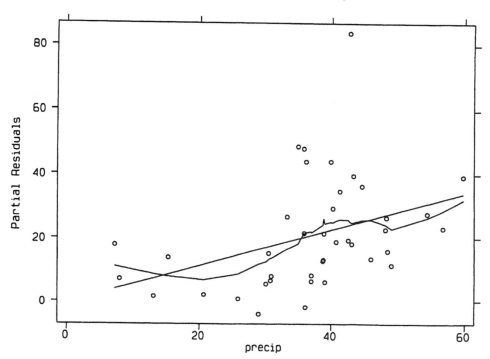

```
fpredict cook, cooksd
list town so2 cook if cook>4/40
```

	town	so2	cook
1.	Phoenix	10	.2543286
24.	Philad	69	.3686437
29.	Provid	94	.2839324

The first instruction stores the Cook's distance statistics in the variable *cook* and the second lists details of those observations for which the statistics is above the suggested cut-off point.

There are three influential observations. Several of the diagnostic procedures used previously also suggest these observation as possibly giving rise to problems and some consideration should be given to repeating the analyses with these three observations removed in addition to the initial removal of *Chicago*.

3.4 Exercises

1. Repeat the analyses described in this chapter after removing the three possible outlying observations identified by the use of the Cook's distance statistic.

2. The solution to the high correlation of the variables *manuf* and *pop* adopted in the chapter was simply to remove the former. Investigate other possibilities such as defining a new variable *manuf/pop* in addition to *pop* to be used in the regression analysis.

3. Consider the possibility of taking a transformation of sulphur dioxide pollution before undertaking any regression analyses. For example, try a *log* transformation.

4. Explore the use of the many other diagnostic procedures available with the fit procedure.

See also exercises in Chapter 12.

CHAPTER 4

Analysis of Variance I: Treating Hypertension

4.1 Description of data

Maxwell and Delaney (1990) describe a study in which the effects of three possible treatments for hypertension were investigated. The details of the treatments are as follows:

Treatment	Description	Levels
drug	medication	drug X, drug Y, drug Z
biofeed	physiological feedback	present, absent
diet	special diet	present, absent

All 12 combinations of the three treatments were included in a $3 \times 2 \times 2$ design. Seventy-two subjects suffering from hypertension were recruited and six were allocated to each combination of treatments. Blood pressure measurements

Table 4.1 Data in *bp.raw*
(Taken from Maxwell and Delaney (1990) with permission from their publisher, Wadsworth)

Biofeed present drug X	Biofeed present drug Y	Biofeed present drug Z	Biofeed absent drug X	Biofeed absent drug Y	Biofeed absent drug Z
Diet absent					
170	186	180	173	189	202
175	194	187	194	194	228
165	201	199	197	217	190
180	215	170	190	206	206
160	219	204	176	199	224
158	209	194	198	195	204
Diet present					
161	164	162	164	171	205
173	166	184	190	173	199
157	159	183	169	196	170
152	182	156	164	199	160
181	187	180	176	180	179
190	174	173	175	203	179

were made on each subject leading to the data shown in Table 4.1. Questions of interest concern differences in mean blood pressure for the different levels of the three treatments and the possibility of interactions between the treatments.

4.2 Analysis of variance model

A suitable model for these data is

$$y_{ijkl} = \mu + \alpha_i + \beta_j + \gamma_k + (\alpha\beta)_{ij} + (\alpha\gamma)_{ik} + (\beta\gamma)_{jk} + (\alpha\beta\gamma)_{ijk} + \epsilon_{ijk} \quad (4.1)$$

where y_{ijkl} represents the blood pressure of the lth subject for the ith drug, the jth level of biofeedback, and the kth level of diet, μ is the overall mean, $\alpha_i, \beta_j,$ and γ_k are the main effects for drugs, biofeedback, and diets, $\alpha\beta, \alpha\gamma,$ and $\beta\gamma$ are the first-order interaction terms, $\alpha\beta\gamma$ a is a second-order interaction term, and ϵ_{ijkl} are the residual or error terms assumed to be normally distributed with zero mean and variance σ^2.

4.3 Analysis using Stata

Assuming the data are in an ASCII file *bp.raw*, exactly as shown in Table 4.1, i.e., 12 rows, the first containing the observations 170 186 180 173 189 202, they may be read into Stata by producing a dictionary file, *bp.dict* containing the following statements:

```
dictionary using bp.raw{
    _column(6) int bp11
    _column(14) int bp12
    _column(22) int bp13
    _column(30) int bp01
    _column(38) int bp02
    _column(46) int bp03
}
```

and using the following command:

```
infile using bp
```

   Note that it was not necesary to define a dictionary here since the same result could have been achieved using a simple `infile` command (see exercises). We now need to reshape the data into a long shape where a single variable, *bp*, contains the blood pressures and the information on treatment condition is contained in three additional variables, *drug*, *biofeed*, and *diet*. The variable *diet* should take on one value for the first six rows and another for the following rows. This is achieved using the commands

```
gen diet=0 if _n<=6
replace diet=1 if _n>6
```

or, more concisely, using

```
gen diet=cond(_n<=6,0,1)
```

We now need to use the *reshape long* command to stack the columns on top of each other. If we specify *bp0* and *bp1* as the variable names in the reshape command, then *bp01*, *bp02*, and *bp03* are stacked into one column with variable name *bp0* (and similarly for *bp1*) and another variable is created that contains the suffixes 1, 2, and 3. We ask for this latter variable to be called *drug* using the option j(drug) as follows:

```
gen id=_n
reshape long bp0 bp1, i(id) j(drug)
list in 1/9
```

	id	drug	diet	bp0	bp1
1.	1	1	0	173	170
2.	1	2	0	189	186
3.	1	3	0	202	180
4.	2	1	0	194	175
5.	2	2	0	194	194
6.	2	3	0	228	187
7.	3	1	0	197	165
8.	3	2	0	217	201
9.	3	3	0	190	199

Here, *id* was generated because we need to specify the row indicator in i(id).

We now need to run the reshape long command again to stack up the columns *bp0* and *bp1* and generate the variable *biofeed*. The instructions to achieve this and to label all the variables are given below.

```
replace id=_n
reshape long bp, i(id) j(biofeed)
replace id=_n

label drop _all
label define d 0 "absent" 1 "present"
label values diet d
label values biofeed d
label define dr 1 "Drug X" 2 "Drug Y" 3 "Drug Z"
label values drug dr
```

To begin, it will be helpful to look at some summary statistics for each of the cells of the design. A simple way of obtaining the required summary measures is to use the table instruction

```
table drug, contents(freq mean bp median bp sd bp)
by(diet biofeed)
```

```
----------+-----------------------------------------------
diet,     |
biofeed   |
and drug  |    Freq.    mean(bp)     med(bp)      sd(bp)
----------+-----------------------------------------------
absent    |
absent    |
  Drug X  |      6         188          192     10.86278
  Drug Y  |      6         200          197     10.07968
  Drug Z  |      6         209          205      14.3527
----------+-----------------------------------------------
absent    |
present   |
  Drug X  |      6         168        167.5     8.602325
  Drug Y  |      6         204          205     12.68069
  Drug Z  |      6         189        190.5     12.61745
----------+-----------------------------------------------
present   |
absent    |
  Drug X  |      6         173          172     9.797959
  Drug Y  |      6         187          188     14.01428
  Drug Z  |      6         182          179      17.1114
----------+-----------------------------------------------
present   |
present   |
  Drug X  |      6         169          167     14.81891
  Drug Y  |      6         172          170     10.93618
  Drug Z  |      6         173        176.5      11.6619
----------+-----------------------------------------------
```

Display 4.1

The standard deviations in Display 4.1 indicate that there are considerable differences in the within cell variability. This might have implications for the analysis of variance of these data since one of the assumptions made is that the observations within each cell have the same variance. To begin, however, apply the model specified in Section 3.2 to the raw data using the anova instruction

```
anova bp drug diet biofeed diet*drug diet*biofeed /*
    */ drug*biofeed drug*diet*biofeed
```

```
                      Number of obs =      72      R-squared      =  0.5840
                      Root MSE      = 12.5167      Adj R-squared =  0.5077

         Source |  Partial SS    df      MS           F      Prob > F
    ------------+--------------------------------------------------------
          Model |   13194.00     11   1199.45455     7.66      0.0000
                |
           drug |    3675.00      2    1837.50       11.73      0.0001
           diet |    5202.00      1    5202.00       33.20 0.0000
        biofeed |    2048.00      1    2048.00       13.07 0.0006
      diet*drug |     903.00      2     451.50        2.88 0.0638
   diet*biofeed |      32.00      1      32.00         .20 0.652
   drug*biofeed |     259.00      2     129.50         .83 0.4425
drug*diet*biofeed |   1075.00      2     537.50        3.43 0.0388
                |
       Residual |    9400.00     60    156.666667

    ------------+--------------------------------------------------------
          Total |   22594.00     71    318.225352
```

The root MSE is simply the square root of the residual mean square, with R-squared and Adj R-squared being as described in Chapter 3. The F-statistic of each effect represents the mean sum of squares for that effect, divided by the residual mean sum of squares, given under the heading MS. The main effects of *drug* ($F_{2,60} = 11.73$, $p < 0.001$), *diet* ($F_{1,60} = 33.20$, $p < 0.001$), and *biofeed* ($F_{1,60} = 13.07$, $p < 0.001$) are all highly significant, and the three-way interaction *drug*diet*biofeed* is also significant beyond the 5% level. The existence of a three-way interaction complicates the interpretation of the other terms in the model; it implies that the interaction between any two of the factors is different at the different levels of the third factor. Perhaps the best way of trying to understand the meaning of the three-way interaction is to plot a number of *interaction diagrams;* that is, plots of mean values for a factor at the different levels of the other factors.

This can be done by first creating a variable *predbp* containing the predicted means (which in this case coincide with the observed cell means because the model is saturated) using the command

```
predict predbp
```

In order to produce a graph with separate lines for the factor *diet*, we need to generate the variables *bp0* and *bp1* as follows:

```
gen bp0=predbp if diet==0
label variable bp0 "diet absent"
gen bp1=predbp if diet==1
label variable bp1 "diet present"
```

Plots of *predbp* against *biofeed* for each level of *drug* with separate lines for *diet* can be obtained using the instructions

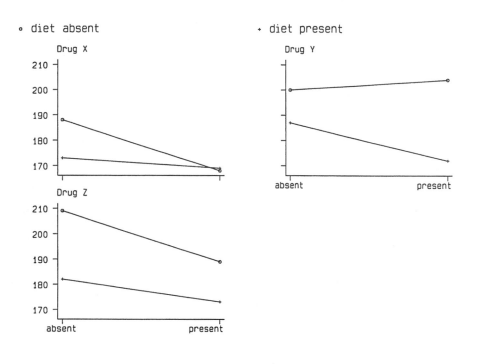

Blood pressure against biofeedback

Figure 4.1 *Interaction diagrams showing the interaction between diet and biofeed for each level of drug.*

```
sort drug
graph bp0 bp1 biofeed, sort c(ll) by(drug) xlab(0,1) ylab /*
    */ b2(" ") b1("Blood pressure against biofeedback ")
```

The resulting interaction diagrams are shown in Figure 4.1. For drug Y, the presence of biofeedback increases the effect of diet; whereas for drug Z, the effect of diet is hardly altered by the presence of biofeedback; and for drug X, the effect is decreased.

Tables of the cell means plotted in the interaction diagrams, as well as the corresponding standard deviations, are produced for each drug using the following command:

```
table diet biofeed, contents(mean bp sd bp) by(drug)
```

```
----------+------------------
drug and  |      biofeed
diet      |   absent    present
----------+------------------
Drug X    |
   absent |      188       168
          | 10.86278  8.602325
          |
  present |      173       169
          | 9.797959  14.81891
----------+------------------
Drug Y    |
   absent |      200       204
          | 10.07968  12.68069
          |
  present |      187       172
          | 14.01428  10.93618
----------+------------------
Drug Z    |
   absent |      209       189
          | 14.3527   12.61745
          |
  present |      182       173
          | 17.1114   11.6619
----------+------------------
```

As mentioned previously, the observations in the 12 cells of the $3 \times 2 \times 2$ design have variances that differ considerably. Consequently, an analysis of variance of the data transformed in some way might be worth considering. For example, to analyze the log transformed observations, we can use the following instructions:

```
gen lbp=log(bp)
anova lbp drug diet biofeed diet*drug diet*biofeed /*
   */ drug*biofeed drug*diet*biofeed
```

```
                     Number of obs =      72   R-squared      =  0.5776
                     Root MSE      = .068013   Adj R-squared  =  0.5002

         Source |  Partial SS    df       MS              F     Prob > F
    ------------+----------------------------------------------------
          Model |  .379534762    11   .03450316           7.46    0.0000
                |
           diet |  .149561559     1   .149561559         32.33    0.0000
           drug |  .107061236     2   .053530618         11.57    0.0001
        biofeed |  .061475507     1   .061475507         13.29    0.0006
      diet*drug |  .024011594     2   .012005797          2.60    0.0830
   diet*biofeed |  .000657678     1   .000657678          0.14    0.7075
   drug*biofeed |  .006467873     2   .003233936          0.70    0.5010
diet*drug*biofeed| .030299315     2   .015149657          3.28    0.0447
                |
       Residual |  .277545987    60   .004625766
    ------------+----------------------------------------------------
          Total |  .657080749    71   .009254658
```

The result is similar to the untransformed blood pressures. Since the three-way interaction is only marginally significant and if no substantive explanation of this interaction is available, it might be better to interpret the results in terms of the very significant main effects. The relevant summary statistics for the log-transformed blood pressures can be obtained using the following instructions:

```
table drug, contents(mean lbp sd lbp)
```

```
----------+------------------------
   drug |   mean(lbp)      sd(lbp)
----------+------------------------
 Drug X |    5.159152      .075955
 Drug Y |    5.247087     .0903675
 Drug Z |    5.232984     .0998921
----------+------------------------
```

```
table diet, contents(mean lbp sd lbp)
```

```
----------+------------------------
   diet |   mean(lbp)      sd(lbp)
----------+------------------------
 absent |    5.258651     .0915982
present |    5.167498     .0781686
----------+------------------------
```

```
table biofeed, contents(mean lbp sd lbp)
```

```
----------+------------------------
biofeed |   mean(lbp)      sd(lbp)
----------+------------------------
 absent |    5.242295     .0890136
present |    5.183854     .0953618
----------+------------------------
```

Drug X appears to produce lower blood pressures, as does the special diet and the presence of biofeedback. Readers are encouraged to try other transformations.

Note that the model with main effects only can also be estimated using regression with dummy variables. Since *drug* has three levels and therefore requires two dummy variables, we save some time by using the xi: prefix as follows:

```
xi: regress lbp i.drug i.diet i.biofeed
```

```
i.drug              Idrug_1-3   (naturally coded; Idrug_1 omitted)
i.diet              Idiet_0-1   (naturally coded; Idiet_0 omitted)
i.biofeed           Ibiofe_0-1  (naturally coded; Ibiofe_0 omitted)

  Source |      SS        df        MS              Number of obs =      72
---------+------------------------------           F( 4,    67) =   15.72
   Model | .318098302      4   .079524576           Prob > F      =  .0000
Residual | .338982447     67   .00505944            R-squared     = 0.4841
---------+------------------------------           Adj R-squared = 0.4533
   Total | .657080749     71   .009254658           Root MSE      = .07113


------------------------------------------------------------------------------
     lbp |    Coef.   Std. Err.        t    P>|t|    [95% Conf. Interval]
---------+--------------------------------------------------------------------
 Idrug_2 |  .0879354   .0205334      4.283   0.000    .0469506    .1289203
 Idrug_3 |  .0738315   .0205334      3.596   0.001    .0328467    .1148163
 Idiet_1 | -.0911536   .0167654     -5.437   0.000   -.1246175   -.0576896
 Ibiofe_1| -.0584406   .0167654     -3.486   0.001   -.0919046   -.0249767
   _cons |  5.233949   .0187443    279.228   0.000    5.196535    5.271363
------------------------------------------------------------------------------
```

The coefficients represent the mean differences between each level compared with the reference level (the omitted categories drug X, diet absent, and biofeed absent) when the other variables are equal to the reference levels. Due to the balanced nature of the design (see next chapter), the p-values are equal to those of ANOVA except that no overall p-value for *drug* is given. This may be obtained using

```
testparm Idrug*
```

```
( 1)  Idrug_2 = 0.0
( 2)  Idrug_3 = 0.0

     F( 2,    67) =    10.58
          Prob > F =     0.0001
```

The F-statistic is different than the one in the last anova command because no interactions were included in the model so that the residual degrees of freedom and the residual sum of squares are both greater than before.

4.4 Exercises

1. Reproduce the result of the command infile using bp without using the dictionary and follow the reshape instructions to generate the required dataset.

2. Produce a diagram that contains boxplots of the observations for each level of drug, for each level of diet, and for each level of biofeedback.

3. Investigate other possible transformations of the data.

4. Suppose that in addition to the blood pressure of each of the individuals in the study, the investigator had also recorded their ages in file *age.dat* with the results shown in Table 4.2. Reanalyze the data using age as a covariate (see **help anova**).

Table 4.2 Ages in *age.dat* to be used as a covariate (see Exercise 4)

id	age	id	age
1	39	37	45
2	39	38	58
3	61	39	61
4	50	40	47
5	51	41	67
6	43	42	49
7	59	43	54
8	50	44	48
9	47	45	46
10	60	46	67
11	77	47	56
12	57	48	54
13	62	49	66
14	44	50	43
15	63	51	47
16	77	52	35
17	56	53	50
18	62	54	60
19	44	55	73
20	61	56	46
21	66	57	59
22	52	58	65
23	53	59	49
24	54	60	52
25	40	61	40
26	62	62	80
27	68	63	46
28	63	64	63
29	47	65	56
30	70	66	58
31	57	67	53
32	51	68	56
33	70	69	64
34	57	70	57
35	64	71	60
36	66	72	48

CHAPTER 5

Analysis of Variance II: Effectiveness of Slimming Clinics

5.1 Description of data

Slimming clinics aim to help people lose weight by offering encouragement and support about dieting through regular meetings. In a study of their effectiveness, a 2×2 factorial design was used to investigate whether giving clients a technical manual containing slimming advice based on psychological behaviorist theory would help them control their diet, and how this might be affected by whether or not a client had already been trying to slim. The data collected are shown in Table 5.1. (They are also given in Hand et al. 1994.) The response variable was defined as follows:

$$\frac{\text{weight at 3 months} - \text{ideal weight}}{\text{natural weight} - \text{ideal weight}} \qquad (5.1)$$

Table 5.1 Data in *slim.dat*

cond	status	resp	cond	status	resp
1	1	−14.67	1	1	−1.85
1	1	−8.55	1	1	−23.03
1	1	11.61	1	2	0.81
1	2	2.38	1	2	2.74
1	2	3.36	1	2	2.10
1	2	−0.83	1	2	−3.05
1	2	−5.98	1	2	−3.64
1	2	−7.38	1	2	−3.60
1	2	−0.94	2	1	−3.39
2	1	−4.00	2	1	−2.31
2	1	−3.60	2	1	−7.69
2	1	−13.92	2	1	−7.64
2	1	−7.59	2	1	−1.62
2	1	−12.21	2	1	−8.85
2	2	5.84	2	2	1.71
2	2	−4.10	2	2	−5.19
2	2	0.00	2	2	−2.80

The number of observations in each cell of the design is not the same, so this is an example of an *unbalanced* 2×2 design.

5.2 Analysis of variance model

A suitable analysis of variance model for the data is

$$y_{ijk} = \mu + \alpha_i + \beta_j + \gamma_{ij} + \epsilon_{ijk} \tag{5.2}$$

where y_{ijk} represents the weight change of the kth individual having status j and condition i, μ is the overall mean, α_i represents the effect of condition i, β_j the effect of status j, γ_{ij} the status × condition interaction, and ϵ_{ijk} the residuals—these are assumed to have a normal distribution with variance σ^2.

The unbalanced nature of the slimming data presents some difficulties for analysis not encountered in factorial designs having the same number of observations in each cell (see previous chapter). The main problem is that when the data are unbalanced, there is no unique way of finding a 'sum of squares' corresponding to each main effect and their interactions, since these effects are no longer independent of one another. If the data were balanced, the total sum of squares would partition orthogonally into three component sums of squares representing the two main effects and their interaction. Several methods have been proposed for dealing with this problem and each leads to a different partition of the overall sum of squares. The different methods for arriving at the sums of squares for unbalanced designs can be explained in terms of the comparisons of different sets of specific models. For a design with two factors A and B, Stata can calculate the following types of sums of squares.

5.2.1 Sequential sums of squares

Sequential sums of squares (also known as hierarchical) represent the effect of adding a term to an existing model. So, for example, a set of sequential sums of squares such as

Source	SS
A	SSA
B	SSB|A
AB	SSAB|A,B

represent a comparison of the following models:

- SSAB|A,B—model including an interaction and main effects compared with one including only main effects.
- SSB|A—model including both main effects, but with no interaction, compared with one including only the main effects of factor A.
- SSA—model containing only the A main effect compared with one containing only the overall mean.

The use of these sums of squares in a series of tables in which the effects are considered in different orders (see later) will often provide the most satisfactory way of answering the question as to which model is most appropriate for the observations. (These are SAS Type I sums of squares; see Everitt and Der, 1996.)

5.2.2 Unique sums of squares

By default, Stata produces unique sums of squares that represent the contribution of each term to a model including all the other terms. So, for a two-factor design, the sums of squares represent the following.

Source	SS
A	SSA|B,AB
B	SSB|A,AB
AB	SSAB|A,B

(These are SAS Type III sums of squares.) Note that these sums of squares generally do not add up to the total sums of squares.

5.2.3 Regression

As we have shown in Chapter 4, ANOVA models can also be estimated using regression by defining suitable dummy variables. Assume that A is represented by a single dummy variable. The regression coefficient for A represents the *partial* contribution of that variable, adjusted for all other variables in the model, say B. This is equivalent to the contribution of A to a model already including B. A complication with regression models is that, in the presence of an interaction, the p-values of the terms depend on the exact coding of the dummy variables (see Aitkin, 1978). The unique sums of squares correspond to regression where dummy variables are coded in a particular way; for example, a two-level factor must be coded as $-1, 1$. For a more detailed explanation of the various types of sums of squares, see Boniface (1995).

There have been numerous discussions over which sums of squares are most appropriate for the analysis of unbalanced designs. The Stata manual appears to recommend its default for general use. Nelder (1977) and Aitkin (1978), however, are strongly critical of 'correcting' main effects for an interaction term involving the same factor; their criticisms are based on both theoretical and pragmatic arguments and seem compelling. A frequently used approach is therefore to test the highest order interaction adjusting for all lower order interactions and not vice versa. Both Nelder and Aitkin prefer the use of Type I sums of squares in association with different orders of effects as the procedure most likely to identify an appropriate model for a dataset.

5.3 Analysis using Stata

The data can be read in from an ASCII file, *slim.dat*, in the usual way using

```
infile cond status resp using slim.dat
```

A table showing the unbalanced nature of the 2×2 design can be obtained from

```
tabulate cond status
```

```
          | status
   cond |       1          2 |    Total
---------+---------------------+----------
     1 |       5         12 |       17
     2 |      11          6 |       17
---------+---------------------+----------
  Total |      16         18 |       34
```

We now use the **anova** command with no options specified to obtain the unique (Type III) sums of squares

```
anova resp cond status cond*status
```

```
               Number of obs =      34    R-squared     =  0.2103
               Root MSE     =  5.9968    Adj R-squared =  0.1313

      Source | Partial SS   df       MS          F      Prob > F
   ---------+----------------------------------------------------
       Model | 287.231861    3  95.7439537      2.66     0.0659
             |
        cond | 2.19850409    1  2.19850409       .06     0.8064
      status | 265.871053    1  265.871053      7.39     0.0108
 cond*status | .130318264    1  .130318264       .00     0.9524
             |
    Residual | 1078.84812   30   35.961604
   ---------+----------------------------------------------------
       Total | 1366.07998   33  41.3963631
```

Our recommendation is that the sums of squares shown in this table are *not* used to draw inferences because the main effects have been adjusted for the interaction.

Instead, the preferred analysis consists of obtaining two sets of sequential sums of squares, the first using the order cond status cond*status and the second the order status cond cond*status; the necessary instructions are

```
anova resp cond status cond*status, sequential
```

```
                       Number of obs =      34      R-squared      =   0.2103
                       Root MSE      = 5.9968      Adj R-squared =   0.1313

        Source |    Seq. SS    df        MS           F     Prob > F
   ------------+-------------------------------------------------------------
         Model |   287.231861    3  95.7439537        2.66     0.0659
               |
          cond |   21.1878098    1  21.1878098        0.59     0.4487
        status |   265.913733    1  265.913733        7.39     0.0108
   cond*status |   .130318264    1  .130318264         .00     0.9524
               |
      Residual |   1078.84812   30  35.961604
   ------------+-------------------------------------------------------------
         Total |   1366.07998   33  41.3963631
```

anova resp status cond cond*status, sequential

```
                       Number of obs =      34      R-squared      =   0.2103
                       Root MSE      = 5.9968      Adj R-squared =   0.1313

        Source |    Seq. SS    df        MS           F     Prob > F
   ------------+-------------------------------------------------------------
         Model |   287.231861    3  95.7439537        2.66     0.0659
               |
        status |   284.971071    1  284.971071        7.92     0.0085
          cond |   2.13047169    1  2.13047169         .06     0.8094
   cond*status |   .130318264    1  .130318264         .00     0.9524
               |
      Residual |   1078.84812   30  35.961604

   ------------+-------------------------------------------------------------
         Total |   1366.07998   33  41.3963631
```

The sums of squares corresponding to model and residuals are, of course, the same in both tables, as is the sum of squares for the interaction term. What differs are the sums of squares in the *cond* and *status* rows in the two tables. The terms of most interest are the sum of squares of status|cond, which is obtained from the table as 265.91, and the sum of squares of cond|status, which is 2.13. These sums of squares differ from the sums of squares for *status* and *cond* alone by 19.06, a portion of the sums of squares that cannot be attributed to either of the variables. The associated F-tests in the two tables make it clear that there is no interaction effect and that status|cond is significant but cond|status is not. The conclusion is that only *status*, i.e., whether or not the woman had been slimming for over 1 year, is important in determining weight change. Provision of the manual appears to have no discernible effect.

Results equivalent to the unique (Type III) sums of squares can be obtained using regression:

```
gen cond1=cond
recode cond1 1=-1 2=1
gen status1=status
```

```
recode status1 1=-1 2=1
gen statcond = cond1*status1
regress res cond1 status1 statcond
```

Source	SS	df	MS		Number of obs =	34
					F(3, 30) =	2.66
Model	287.231861	3	95.7439537		Prob > F =	0.0659
Residual	1078.84812	30	35.961604		R-squared =	0.2103
					Adj R-squared =	0.1313
Total	1366.07998	33	41.3963631		Root MSE =	5.9968

resp	Coef.	Std. Err.	t	P>|t|	[95% Conf. Interval]	
cond1	.2726251	1.102609	0.247	0.806	-1.979204	2.524454
status1	2.998042	1.102609	2.719	0.011	.746213	5.24987
statcond	-.066375	1.102609	-0.060	0.952	-2.318204	2.185454
_cons	-3.960958	1.102609	-3.592	0.001	-6.212787	-1.70913

These results differ from the regression used by Stata's anova with the option regress:

```
anova resp cond status cond*status, regress
```

Source	SS	df	MS		Number of obs =	34
					F(3, 30) =	2.66
Model	287.231861	3	95.7439537		Prob > F =	0.0659
Residual	1078.84812	30	35.961604		R-squared =	0.2103
					Adj R-squared =	0.1313
Total	1366.07998	33	41.3963631		Root MSE =	5.9968

resp		Coef.	Std. Err.	t	P>|t|	[95% Conf. Interval]	
_cons		-.7566666	2.448183	-.309	0.759	-5.756524	4.24319
cond							
	1	-.4125001	2.9984	-.138	0.891	-6.536049	5.711049
	2	(dropped)					
status							
	1	-5.863333	3.043491	-1.927	0.064	-12.07897	.3523044
	2	(dropped)					
cond*status							
1	1	-.2655002	4.410437	-.060	0.952	-9.272815	8.741815
1	2	(dropped)					
2	1	(dropped)					
2	2	(dropped)					

because this uses different dummy variables. The dummy variables are equal to 1 for the levels to the left of the reported coefficient and zero otherwise; that is, the dummy variable for cond*status is 1 when *status* and *cond* are both 1. A table of mean values helpful in interpreting these results can be found using

```
table cond status, c(mean resp) row col f(%8.2f)
```

```
----------+--------------------
          |        status
   cond   |     1        2  Total
----------+--------------------
      1  | -7.30   -1.17   -2.97
      2  | -6.62    -.76   -4.55
         |
   Total | -6.83   -1.03   -3.76
----------+--------------------
```

The means demonstrate that experienced slimmers achieve the greatest weight reduction.

5.4 Exercises

1. Investigate what happens to the sequential sums of squares if the cond*status interaction term is given before the main effects cond status in the anova command with the sequential option.

2. Use fit or regress to reproduce the analysis of variance by coding both condition and status as (0,1) dummy variables and creating an interaction variable as the product of these dummy variables.

3. Use regress in conjunction with xi: to fit the same model without the need to generate any dummy variables (see help xi:).

4. Reproduce the results of anova resp cond status cond*status, regress using regress by making xi: omit the last category instead of the first (see help xi:).

See also the exercises in Chapters 7 and 13.

CHAPTER 6

Logistic Regression: Treatment of Lung Cancer and Diagnosis of Heart Attacks

6.1 Description of data

Two datasets will be analyzed in this chapter. The first dataset shown in Table 6.1 originates from a clinical trial in which lung cancer patients were randomized to receive two different kinds of chemotherapy (sequential therapy and alternating therapy). The outcome was classified into one of four categories: progressive disease, no change, partial remission, or complete remission. The data were published in Holtbrugge and Schumacher (1991) and also appear in Hand et al. (1994). The main aim of any analysis will be to assess differences between the two therapies.

Table 6.1 Tumor data

Therapy	Sex	Progressive disease	No change	Partial remission	Complete remission
Sequential	Male	28	45	29	26
	Female	4	12	5	2
Alternative	Male	41	44	20	20
	Female	12	7	3	1

The second dataset arises from a study to investigate the use of serum creatine kinase (CK) levels for the diagnosis of myocardial infarction (heart attack). Patients admitted to a coronary care unit because they were suspected of having had a myocardial infarction within the last 48 hours had their CK levels measured on admission and the next two mornings. A clinician who was "blind" to the CK results came to an independent "gold standard" diagnosis using electrocardiograms, clinical records, and autopsy reports. The maximum CK levels for 360 patients are given in Table 6.2, together with the clinician's diagnosis. The table was taken from Sackett et al. (1991), where only the ranges of CK levels were given, not their precise values.

The main question of interest here is how well CK discriminates between those with and without myocardial infarction and to investigate the characteristics of the diagnostic test for different thresholds.

Table 6.2 Diagnosis of myocardial infarct from serum creatine kinase (CK)
(Taken from Sacket et al. (1991) with permission of the publisher, Little Brown &
Company)

Maximum CK level	Infarct present	Infarct absent
0-39	2	88
40-79	13	26
80-119	30	8
120-159	30	5
160-199	21	0
200-239	19	1
240-279	18	1
280-319	13	1
320-359	19	0
360-399	15	0
400-439	7	0
440-479	8	0
480-	35	0

6.2 The logistic regression model

Logistic regression is used when the response variable is dichotomous. In this
case, we are interested in how the probability that the response variable takes
on the value of interest (usually coded as 1, the other value being zero) depends
on a number of explanatory variables. The probability π that $y = 1$ is just the
expected value of y. In linear regression (see Chapter 2), the expected value of
y is modeled as a linear function of the explanatory variables

$$E[y] = \beta_0 + \beta_1 x_1 + \beta_2 x_2 + \cdots + \beta_p x_p \qquad (6.1)$$

However, there are two problems with using the method of linear regression
when y is dichotomous: (1) the predicted probability must satisfy $0 \leq \pi \leq
1$, whereas a linear predictor can yield any value from minus infinity to plus
infinity, and (2) the observed values of y do not follow a normal distribution
with mean π, but rather a Bernoulli (or Binomial(1,π)) distribution.

In logistic regression, the first problem is addressed by replacing the probability $\pi = E[y]$ on the left-hand side of equation (6.1) by the *logit* of the
probability, giving

$$\text{logit}(\pi) = \log(\pi/(1 - \pi)) = \beta_0 + \beta_1 x_1 + \beta_2 x_2 + \cdots + \beta_p x_p \qquad (6.2)$$

The logit of the probability is simply log of the odds of the event of interest.
Writing $\boldsymbol{\beta}$ and $\mathbf{x}_i$ for the column vectors $(\beta_0, \cdots, \beta_p)^T$ and $(1, x_{1i}, \cdots, x_{pi})^T$,

respectively, the predicted probability as a function of the linear predictor is

$$\pi\left(\boldsymbol{\beta}^T \mathbf{x}_i\right) = \frac{\exp\left(\boldsymbol{\beta}^T \mathbf{x}_i\right)}{1 + \exp\left(\boldsymbol{\beta}^T \mathbf{x}_i\right)}$$

$$= \frac{1}{1 + \exp\left(-\boldsymbol{\beta}^T \mathbf{x}_i\right)} \tag{6.3}$$

When the logit takes on any real value, this probability always satisfies $0 \leq \pi(\boldsymbol{\beta}^T \mathbf{x}_i) \leq 1$.

The second problem relates to the estimation procedure. Whereas maximum likelihood estimation in linear regression leads to (weighted) least squares, this is not the case in logistic regression. The log likelihood function for logistic regression is

$$l(\boldsymbol{\beta}; \mathbf{y}) = \sum_i \left[y_i \log\left(\pi\left(\boldsymbol{\beta}^T \mathbf{x}_i\right)\right) + (1 - y_i) \log\left(1 - \pi\left(\boldsymbol{\beta}^T \mathbf{x}_j\right)\right) \right] \tag{6.4}$$

This log likelihood is maximized numerically using an iterative algorithm. For full details of logistic regression, see, for example, Collett (1991).

Logistic regression can be generalized to the situation where the response variable has more than two ordered response categories $y_1, \cdots, y_I$ by thinking of these categories as resulting from thresholding an unobserved continuous variable u at a number of cut-points κ_j, $j = 1, \cdots, I - 1$ so that $y = y_1$ if $u \leq \kappa_1$, $y = y_2$ if $\kappa_1 < u \leq \kappa_2$, ..., and $y = y_I$ if $\kappa_{I-1} < u$. The variable $u - \boldsymbol{\beta}^T \mathbf{x}$ is assumed to have the standard logistic distribution $\Pr(u - \boldsymbol{\beta}^T \mathbf{x} < X) = 1/(1 + \exp(-X))$ so that the cumulative probability γ_j of a response up to and including y_j is

$$\gamma_j = \Pr(u \leq \kappa_j) = \Pr\left(u - \boldsymbol{\beta}^T \mathbf{x} \leq \kappa_j - \boldsymbol{\beta}^T \mathbf{x}\right)$$

$$= \frac{1}{1 + \exp\left(\boldsymbol{\beta}^T \mathbf{x} - \kappa_j\right)} \tag{6.5}$$

where κ_I is taken to be ∞. The probability of the jth response category is then simply $\pi_j = \gamma_j - \gamma_{j-1}$. Ordinary logistic regression is a special case of the ordinal regression model where $I = 2$, the cut point is $\kappa_1 = 0$, and the probability of interest is $\pi_2 = 1 - \gamma_1$ giving the result in (6.3). The ordinal regression model is called the *proportional odds* model because the log odds that $y \leq y_j$ is

$$\log\left(\frac{\gamma_j(x)}{1 - \gamma_j(x)}\right) = \kappa_j - \boldsymbol{\beta}^T \mathbf{x} \tag{6.6}$$

so that the log odds ratio for two values of $\mathbf{x}$ is $\boldsymbol{\beta}^T(\mathbf{x}_1 - \mathbf{x}_2)$ and is independent of j.

Note that the probit and ordinal probit models correspond to logistic and or-

dinal logistic regression models with the cumulative distribution function in
equation (6.5) replaced by the standard normal distribution.

6.3 Analysis using Stata

6.3.1 Chemotherapy treatment of lung cancer

Assume the ASCII file *tumor.dat* contains the four-by-four matrix of frequencies
shown in Table 6.1. First read the data and generate indicator variables for
therapy and sex:

```
infile fr1 fr2 fr3 fr4 using tumour.dat
gen therapy=int((_n-1)/2)
sort therapy
by therapy: gen sex=_n
label define t 0 seq 1 alt
label values therapy t
label define s 1 male 2 female
label values sex s
```

and then reshape the data to long, placing the four levels of the outcome into
a variable *outc*, and expand the dataset to have one observation per subject:

```
reshape long fr, i(therapy sex) j(outc)
expand fr
```

We can check that the data conversion is correct by tabulating these data as
in Table 6.1.

```
table sex outc, contents(freq) by(therapy)
```

```
----------+------------------------
therapy   |           outc
and sex   |    1     2     3     4
----------+------------------------
seq       |
    male  |   28    45    29    26
  female  |    4    12     5     2
----------+------------------------
alt       |
    male  |   41    44    20    20
  female  |   12     7     3     1
----------+------------------------
```

In order to be able to carry out ordinary logistic regression, we need to
dichotomize the outcome, for example, by considering partial and complete
remission to be an improvement and the other categories as no improvement.
The new outcome variable can be generated as follows:

```
gen improve=outc
recode improve 1/2=0 3/4=1
```

The command `logit` for logistic regression follows the same syntax as `regress` and all other estimation commands. For example, automatic selection procedures can be carried out using `sw`, and post-estimation commands such as `testparm` are available. First, include *therapy* as the only explanatory variable:

```
logit improve therapy
```

```
Iteration 0:  Log Likelihood =-194.40888
Iteration 1:  Log Likelihood =-192.30753
Iteration 2:  Log Likelihood =-192.30471

Logit Estimates                              Number of obs =     299
                                             chi2(1)       =    4.21
                                             Prob > chi2   = 0.0402
Log Likelihood = -192.30471                  Pseudo R2     = 0.0108


-------------------------------------------------------------------------
improve |     Coef.   Std. Err.       z     P>|z|    [95% Conf. Interval]
--------+----------------------------------------------------------------
therapy |  -.4986993   .2443508   -2.041   0.041    -.977618    -.0197805
  _cons |   -.361502   .1654236   -2.185   0.029   -.6857263    -.0372777
-------------------------------------------------------------------------
```

The algorithm takes three iterations to converge. The coefficient of *therapy* represents the difference in the log odds (of an improvement) between the alternating and sequential therapies. The negative value indicates that sequential therapy is superior to alternating therapy. The p-value of the coefficient is 0.041 in the table. This was derived from the Wald statistic, z, which is equal to the coefficient divided by its asymptotic standard error (`Std. Err.`) as derived from the Hessian matrix of the log likelihood function, evaluated at the maximum likelihood solution. This p-value is less reliable than the p-value based on the likelihood ratio between the model including only the constant and the current model, which is given at the top of the output (`chi2(1)=4.21`). Here, minus twice the log of the likelihood ratio is equal to 4.21 which has an approximate χ^2-distribution with one degree of freedom (because there is one additional parameter) giving a p-value of 0.040, very similar to that based on the Wald-test. The coefficient of therapy represents the difference in log odds between the therapies and is not easy to interpret apart from the sign. Taking the exponential of the coefficient gives the odds ratio and exponentiating the 95% confidence limits gives the confidence interval for the odds ratio. Fortunately, the command `logistic` can be used to obtain the required odds ratio and its confidence interval

```
logistic improve therapy
```

```
Logit Estimates                                    Number of obs =      299
                                                   chi2(1)       =     4.21
                                                   Prob > chi2   = 0.0402
Log Likelihood = -192.30471                        Pseudo R2     = 0.0108


-------------------------------------------------------------------------
improve | Odds Ratio  Std. Err.       z     P>|z|    [95% Conf. Interval]
--------+----------------------------------------------------------------
therapy |   .6073201   .1483991   -2.041    .041     .3762061    .9804138
-------------------------------------------------------------------------
```

The standard error now represents the approximate standard error of the odds
ratio (calculated using the delta method). However, the Wald statistic and
confidence interval are derived using the log odds, and its standard error as
the sampling distribution of the log odds is much closer to normal than that of
the odds. In order to be able to test whether the inclusion of *sex* in the model
significantly increases the likelihood, the current likelihood can be saved using

    lrtest, saving(1)

Including sex gives

    logistic improve therapy sex

```
Logit Estimates                                    Number of obs =      299
                                                   chi2(2)       =     7.55
                                                   Prob > chi2   = 0.0229
Log Likelihood = -190.63171                        Pseudo R2     = 0.0194


-------------------------------------------------------------------------
improve | Odds Ratio  Std. Err.       z     P>|z|    [95% Conf. Interval]
--------+----------------------------------------------------------------
therapy |   .6051969   .1486907   -2.044   0.041     .3739084    .9795537
    sex |   .5197993   .1930918   -1.761   0.078     .2509785   1.076551
-------------------------------------------------------------------------
```

and a more reliable p-value is obtained using

    lrtest, saving(0)
    lrtest, model(1) using(0)

```
Logistic:  likelihood-ratio test          chi2(1)    =      3.35
                                          Prob > chi2 =    0.0674
```

which is not very different from the value of 0.078. Note that model(1) refers to
the model that is nested in the more complicated model referred to by using(0).
If the two models had been fitted in the reverse order, it would not have been
necessary to specify these options because lrtest assumes that the current
model is nested in the last model saved using lrtest, saving(0). Retaining
the variable *sex* in the model, the predicted probabilities can be obtained using

```
predict lodds
gen prob=exp(lodds)/(1+exp(lodds))
```

or more simply, using a special prediction command designed for logistic regression,

```
lpredict prob
```

and the four different predicted probabilities can be compared with the observed
proportions as follows:

```
table sex , contents(mean prob mean improve freq) by(therapy)
```

```
----------+-------------------------------------------------
therapy   |
and sex   |    mean(prob)  mean(improve)             Freq.
----------+-------------------------------------------------
seq       |
    male  |     .4332747      .4296875                 128
  female  |     .2843846      .3043478                  23
----------+-------------------------------------------------
alt       |
    male  |     .3163268           .32                 125
  female  |     .1938763       .173913                  23
----------+-------------------------------------------------
```

The agreement is good, so there appears to be no strong interaction between sex
and type of therapy. (We could test for an interaction between sex and therapy
by using xi: logistic improve i.therapy*i.sex). Residuals are not very
informative for these data because there are only four different predicted probabilities.

We now fit the proportional odds model using the full ordinal response variable *outc*:

```
ologit outc therapy sex, table
```

The results are shown in Display 6.1. Both *therapy* and *sex* are more significant
than before. The option table has produced the last part of the output to
remind us how the proportional odds model is defined. We could calculate the
probability that a male (*sex*=1) who is receiving sequential therapy (*therapy*=0)
will be in complete remission (*outc*=4) using

```
display 1-1/(1+exp(-0.5413938-0.758662))
```

```
.21415563
```

but a much quicker way of computing the predicted probabilities for all four
responses and all combinations of explanatory variables is to use the command
ologitp

```
ologitp p1 p2 p3 p4
```

```
Iteration 0:  Log Likelihood =-399.98398
Iteration 1:  Log Likelihood =-394.53988
Iteration 2:  Log Likelihood =-394.52832

Ordered Logit Estimates                              Number of obs =     299
                                                     chi2(2)       =   10.91
                                                     Prob > chi2   = 0.0043
Log Likelihood = -394.52832                          Pseudo R2     = 0.0136


------------------------------------------------------------------------------
    outc |     Coef.   Std. Err.      z     P>|z|      [95% Conf. Interval]
---------+--------------------------------------------------------------------
 therapy |  -.580685   .2121432    -2.737   0.006      -.9964781    -.164892
     sex |  -.5413938  .2871764    -1.885   0.059      -1.104249    .0214616
---------+--------------------------------------------------------------------
   _cut1 |  -1.859437  .3828641              (Ancillary parameters)
   _cut2 |  -.2921603  .3672626
   _cut3 |   .758662   .3741486
------------------------------------------------------------------------------


    outc |      Probability              Observed
---------|------------------------------------------
       1 |   Pr(       xb+u<_cut1)         0.2843
       2 |   Pr(_cut1<xb+u<_cut2)          0.3612
       3 |   Pr(_cut2<xb+u<_cut3)          0.1906
       4 |   Pr(_cut3<xb+u)                0.1639
```

Display 6.1

and to tabulate the results as follows:

```
table sex, contents(mean p1 mean p2 mean p3 mean p4) by(therapy)
```

```
----------+------------------------------------------------
therapy   |
and sex   |   mean(p1)    mean(p2)    mean(p3)    mean(p4)
----------+------------------------------------------------
seq       |
    male  |   .2111441    .3508438    .2238566    .2141556
  female  |   .3150425    .3729235    .175154     .1368799
----------+------------------------------------------------
alt       |
    male  |   .3235821    .3727556    .1713585    .1323038
  female  |   .4511651    .346427     .1209076    .0815003
----------+------------------------------------------------
```


6.3.2 Diagnosis of heart attacks

The data in *sck.dat* are read in using

```
infile ck pres abs using sck.dat
```

Each observation represents all subjects with maximum creatine kinase values in the same range. The total number of subjects is *pres+abs*

```
gen tot=pres+abs
```

and the number of subjects with the disease is *pres*. The probability associated with each observation is Binomial(tot, π). The programs logit and logistic are for data where each observation represents a single Bernouilli trial, with probability Binomial(1, π). Another program, blogit, can be used to analyze the "grouped" data with "denominators" *tot*:

```
blogit pres tot ck
```

```
Logit Estimates                          Number of obs =      360
                                         chi2(1)       = 283.15
                                         Prob > chi2   = 0.0000
Log Likelihood = -93.886407              Pseudo R2     = 0.6013

-----------------------------------------------------------------------
_outcome |     Coef.   Std. Err.      z     P>|z|     [95% Conf. Interval]
---------+-------------------------------------------------------------
      ck |   .0351044   .0040812    8.601   0.000     .0271053    .0431035
   _cons |  -2.326272   .2993611   -7.771   0.000    -2.913009   -1.739535
-----------------------------------------------------------------------
```

There is a very significant association between CK and the probability of infarct. We now need to investigate whether it is reasonable to assume that the log odds depend linearly on CK. Therefore, plot the observed proportions and predicted probabilities as follows:

```
gen prop = pres/tot
predict pred
label variable prop "observed"
label variable pred "predicted"
graph pred prop ck, c(s.) xlab ylab ll("Probability") gap(4)
```

where the option c(s.) causes the first set of points to be connected by a smooth curve. In the resulting graph in Figure 6.1, the curve fits the data reasonably well, the largest discrepency being at CK = 280.

There are far more "post-estimation" commands available for logistic than there are for blogit. Therefore, we transform the data into the form required for logistic:

```
expand tot
sort ck
gen infct=0
quietly by ck: replace infct=1 if _n<=pres
```

Reproducing the results of blogit using logit gives

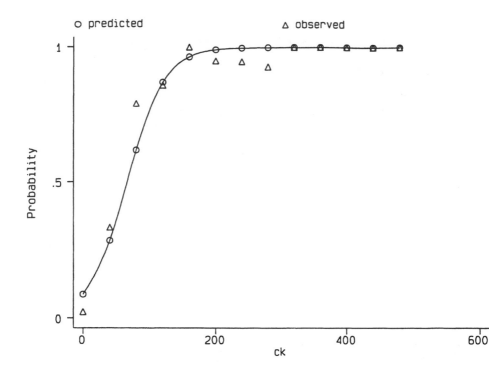

Figure 6.1 *Probability of infarct as a function of creatine kinase levels.*

```
logit infct ck, nolog
```

```
Logit Estimates                               Number of obs =     360
                                              chi2(1)       =  283.15
                                              Prob > chi2   =  0.0000
Log Likelihood = -93.886407                   Pseudo R2     =  0.6013


---------------------------------------------------------------------
  infct |      Coef.   Std. Err.       z     P>|z|    [95% Conf. Interval]
--------+------------------------------------------------------------
     ck |    .0351044   .0040812     8.601   0.000     .0271053   .0431035
  _cons |   -2.326272   .2993611    -7.771   0.000    -2.913009  -1.739535
---------------------------------------------------------------------
```

where the `nolog` option was used to stop the iteration history being given.

In order to compute residuals using the command `lpredict`, we need to run `logistic`, which can be done without producing any output by using `quietly`

```
quietly logistic infct ck
```

One useful type of residual is the Pearson residual for each "covariate pattern," i.e., for each combination of values in the covariates (here for each value of CK). These residuals can be obtained and plotted as follows:

```
lpredict resi, rstandard
graph resi ck, s([ck]) psize(150) xlab ylab gap(5)
```

The graph is shown in Figure 6.2. There are three large outliers. The largest outlier at CK = 280 is due to one subject out of 14 not having had an infarct although the predicted probability of an infarct is almost 1.

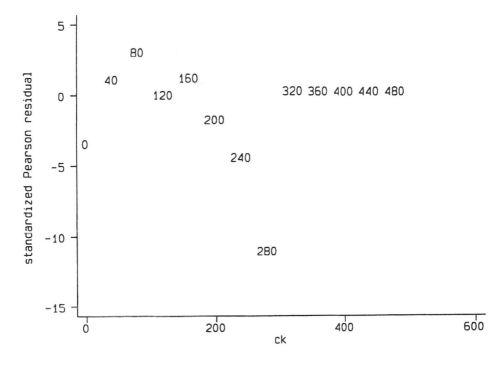

Figure 6.2 *Standardized Pearson residuals vs. creatine kinase level.*

Now we wish to determine the accuracy of the diagnostic test. A *classification table* of the predicted diagnosis (using a cut-off of the predicted probability of 0.5) versus the true diagnosis can be obtained using

```
lstat
```

```
Logistic model for infct

                  -------- True --------
Classified |        D              ~D           Total
-----------+------------------------------+-----------
    +      |       215             16 |          231
    -      |        15            114 |          129
-----------+------------------------------+-----------
  Total    |       230            130 |          360

Classified + if predicted Pr(D) >= .5
True D defined as infct ~= 0


-------------------------------------------------------
Sensitivity                         Pr( +| D)    93.48%
Specificity                         Pr( -|~D)    87.69%
Positive predictive value           Pr( D| +)    93.07%
Negative predictive value           Pr(~D| -)    88.37%
-------------------------------------------------------
False + rate for true ~D            Pr( +|~D)    12.31%
False - rate for true D             Pr( -| D)     6.52%
False + rate for classified +       Pr(~D| +)     6.93%
False - rate for classified -       Pr( D| -)    11.63%
-------------------------------------------------------
Correctly classified                             91.39%
-------------------------------------------------------
```

Both the sensitivity and specificity are relatively high. These characteristics are generally assumed to generalize to other populations, whereas the positive and negative predictive values depend on the prevalence (or prior probability) of the condition (see, for example, Sackett et al., 1991). The CK-value corresponding to a the predicted probability of 0.5 used as a threshold above can be found by setting the linear predictor (the log odds) to zero and solving for CK:

```
disp -_b[_cons]/_b[ck]
```

$$\boxed{66.267255}$$

Since the lower limit of the range of values represented by each category of CK was used, we should add 39/2 to the threshold, giving 85.8.

The use of other probability cut-offs could be investigated using the option cutoff(#) in the above command or using the commands lroc to plot a ROC-curve (sensitivity vs. 1-specificity for different cut-offs) or lsens to plot sensitivity and specificity against cut-off (see exercises).

The above classification table can be misleading because we are testing the model on the same sample that was used to derive it. An alternative approach is to compute predicted probabilities for each observation from a model fitted to the remaining observations. This method, called "leave one out" method or *jacknifing* (see Lachenbruch and Mickey, 1986), can be carried out relatively easily for our data because we only have a small number of covariate and response patterns. First, label each unique covariate pattern consecutively using a variable *num*:

```
lpredict num, number
```

Now define *grp* to label each group of unique covariate and response patterns using

```
sort num infct
gen nxt=0
quietly by num infct:replace nxt=1 if _n==1
gen grp=sum(nxt)
summ grp
```

Variable	Obs	Mean	Std. Dev.	Min	Max
grp	360	8.658333	6.625051	1	20

There are therefore 20 groups of different combinations of CK and true diagnosis. For the jacknifing procedure, run `logistic` 20 times (for each *grp*), excluding one observation from *grp* to derive the model for predicting the probability for all observations in *grp*.

This can be done using the following instructions:

```
replace nxt=grp if nxt==1
gen prp=0
for 1-20,ltype(numeric): /*
     */ logistic infct ck if nxt~=@// predict p /*
     */ // replace prp=p if grp==@// drop p
```

Here, // was used to run four commands in each iteration of the loop to:

1. Derive the model excluding one observation from *grp*

2. Obtain the predicted probabilities *p* (use `predict` because `lpredict` only produces results for the estimation sample)

3. Set *prp* to the predicted probability for all observations in *grp*

4. Drop *p* so that it can be defined again in the next iteration

The classification table for the jacknifed probabilities can be obtained using

```
gen class=cond(prp>=0.5, 1, 0)
tab class infct
```

class	infct 0	1	Total
0	114	15	129
1	16	215	231
Total	130	230	360

giving the same result as before (this is not generally the case).

6.4 Exercises

1. Read in the data without using the `expand` command and reproduce the result of ordinal logistic regressions using the appropriate weights (see help for `ologit`).

2. Test for an association between *depress* and *life* for the data described in Chapter 2 using

   a. Ordinal logistic regression with *depress* as the dependent variable.

   b. Logistic regression with *life* as the dependent variable.

3. Use `sw` together with `logistic` to find a model for predicting *life* using the data from Chapter 2 with different sets of candidate variables (see Chapter 3).

4. Produce a graph similar to that in Figure 6.1 using probit analysis (see help for `bprobit`).

5. Explore the use of `lstat`, `cutoff(#)`, `lroc`, and `lsens` for the diagnosis data.

CHAPTER 7

Generalized Linear Models: Australian School Children

7.1 Description of data

This chapter reanalyzes a number of datasets discussed in previous chapters and, in addition, describes the analysis of a new dataset given in Aitkin (1978). These data come from a sociological study of Australian Aboriginal and white children. The sample included children from four age groups (final year in primary school and first three years in secondary school) who were classified as slow or average learners. The number of days absent from school during the school year was recorded for each child. The data are given in Table 7.1. The variables are as follows:

- *eth:* ethnic group (A=aboriginal, N=white)
- *sex:* sex (M=male, F=female)
- *age:* class in school (F0, F1, F2, F3)
- *lrn:* average or slow learner(SL=slow learner, AL=average learner)
- *days:* number of days absent from school in 1 year

7.2 Generalized linear models

Previous chapters have described analysis of variance and linear and logistic regression. Analysis of variance can be expressed as linear regression by defining appropriate dummy variables. This chapter will describe an even more general class of models, called generalized linear models, of which linear regression and logistic regression are special cases.

Both linear and logistic regression involve a linear combination of the explanatory variables, called the linear predictor, of the form

$$\eta = \beta_0 + \beta x_1 + \beta x_2 + \cdots + \beta x_p$$
$$= \beta^T \mathbf{x} \tag{7.1}$$

In both types of regression, the linear predictor determines the expectation μ of the response variable. In linear regression, where the response is continuous, μ is directly equated with the linear predictor. This is not advisable when the response is dichotomous because in this case the expectation is a probability

Table 7.1 Data in *quine.dta* presented in four columns to save space; the variables are listed in the order *eth, sex, age, lrn, days* (Taken from Aitkin (1978) with permission of the Royal Statistical Society)

1					2					3					4				
A	M	F0	SL	2	A	M	F0	SL	11	A	M	F0	SL	14	A	M	F0	SL	5
A	M	F0	AL	5	A	M	F0	AL	13	A	M	F0	AL	20	A	M	F0	AL	22
A	M	F1	SL	6	A	M	F1	SL	6	A	M	F1	SL	15	A	M	F1	SL	7
A	M	F1	AL	14	A	M	F1	AL	6	A	M	F1	AL	32	A	M	F1	AL	53
A	M	F2	SL	57	A	M	F2	SL	14	A	M	F2	SL	16	A	M	F2	SL	16
A	M	F2	AL	17	A	M	F2	AL	40	A	M	F2	AL	43	A	M	F2	AL	46
A	M	F3	SL	8	A	M	F3	SL	23	A	M	F3	SL	23	A	M	F3	SL	28
A	M	F3	AL	34	A	M	F3	AL	36	A	M	F3	AL	38	A	M	F3	AL	3
A	F	F0	SL	5	A	F	F0	SL	11	A	F	F0	SL	24	A	F	F0	SL	45
A	F	F0	AL	5	A	F	F0	AL	6	A	F	F0	AL	6	A	F	F0	AL	9
A	F	F1	SL	13	A	F	F1	SL	23	A	F	F1	SL	25	A	F	F1	SL	32
A	F	F1	AL	53	A	F	F1	AL	54	A	F	F1	AL	5	A	F	F1	AL	5
A	F	F1	SL	11	A	F	F1	SL	17	A	F	F1	SL	19	A	F	F1	SL	8
A	F	F1	AL	13	A	F	F1	AL	14	A	F	F1	AL	20	A	F	F1	AL	47
A	F	F2	SL	48	A	F	F2	SL	60	A	F	F2	SL	81	A	F	F2	SL	2
A	F	F2	AL	0	A	F	F2	AL	2	A	F	F2	AL	3	A	F	F2	AL	5
A	F	F3	SL	10	A	F	F3	SL	14	A	F	F3	SL	21	A	F	F3	SL	36
A	F	F3	AL	40	A	F	F3	AL	6	A	F	F3	AL	17	A	F	F3	AL	67
N	M	F0	SL	0	N	M	F0	SL	0	N	M	F0	SL	2	N	M	F0	SL	7
N	M	F0	AL	11	N	M	F0	AL	12	N	M	F0	AL	0	N	M	F0	AL	0
N	M	F1	SL	5	N	M	F1	SL	5	N	M	F1	SL	5	N	M	F1	SL	11
N	M	F1	AL	17	N	M	F1	AL	3	N	M	F1	AL	4	N	M	F1	AL	22
N	M	F2	SL	30	N	M	F2	SL	36	N	M	F2	SL	8	N	M	F2	SL	0
N	M	F2	AL	1	N	M	F2	AL	5	N	M	F2	AL	7	N	M	F2	AL	16
N	M	F3	SL	27	N	M	F3	SL	0	N	M	F3	SL	30	N	M	F3	SL	10
N	M	F3	AL	14	N	M	F3	AL	27	N	M	F3	AL	41	N	M	F3	AL	69
N	F	F0	SL	25	N	F	F0	SL	10	N	F	F0	SL	11	N	F	F0	SL	20
N	F	F0	AL	33	N	F	F0	AL	5	N	F	F0	AL	7	N	F	F0	AL	0
N	F	F1	SL	1	N	F	F1	SL	5	N	F	F1	SL	5	N	F	F1	SL	5
N	F	F1	AL	5	N	F	F1	AL	7	N	F	F1	AL	11	N	F	F1	AL	15
N	F	F1	SL	5	N	F	F1	SL	14	N	F	F1	SL	6	N	F	F1	SL	6
N	F	F1	AL	7	N	F	F1	AL	28	N	F	F1	AL	0	N	F	F1	AL	5
N	F	F2	SL	14	N	F	F2	SL	2	N	F	F2	SL	2	N	F	F2	SL	3
N	F	F2	AL	8	N	F	F2	AL	10	N	F	F2	AL	12	N	F	F2	AL	1
N	F	F3	SL	1	N	F	F3	SL	9	N	F	F3	SL	22	N	F	F3	SL	3
N	F	F3	AL	3	N	F	F3	AL	5	N	F	F3	AL	15	N	F	F3	AL	18
N	F	F3	AL	22	N	F	F3	AL	37										

which must satisfy $0 \leq \mu \leq 1$. In logistic regression, the linear predictor is therefore equated with a function of μ, the logit, $\eta = \log(\mu/(1-\mu))$. In generalized linear models, the linear predictor can be equated with any of a number of different functions $g(\mu)$ of μ, called *link functions*; that is,

$$\eta = g(\mu) \tag{7.2}$$

In linear regression, the probability distribution of the response variable is assumed to be normal with mean μ. In logistic regression, a binomial distribution is assumed with probability parameter μ. Both distributions, the normal and binomial distributions, come from the same family of distributions, called the exponential family:

$$f(y; \theta, \phi) = \exp\left\{(y\theta - b(\theta))/a(\phi) + c(y, \phi)\right\} \tag{7.3}$$

For example, for the normal distribution,

$$
\begin{aligned}
f(y; \theta, \phi) &= \frac{1}{\sqrt{(2\pi\sigma^2)}} \exp\left\{-(y-\mu)^2/2\sigma^2\right\} \\
&= \exp\left\{\left(y\mu - \mu^2/2\right)/\sigma^2 - \frac{1}{2}\left(y^2/\sigma^2 + \log\left(2\pi\sigma^2\right)\right)\right\} \tag{7.4}
\end{aligned}
$$

so that $\theta = \mu$, $b(\theta) = \theta^2/2$, $\phi = \sigma^2$ and $a(\phi) = \phi$.

The parameter θ, a function of μ, is called the *canonical link*. The canonical link is frequently chosen as the link function (and is the default link in the Stata command for fitting generalized linear models, glm) although the canonical link is not necessarily more appropriate than any other link. Table 7.2 lists some of the most common distributions and link functions used in generalized linear models.

Table 7.2 Probability distributions and their canonical link functions

Distribution	Variance function	Dispersion parameter	Link function	$g(\mu) = \theta(\mu)$	range of μ for $-\infty \leq g(\mu) \leq \infty$
Normal	1	σ^2	identity	μ	$-\infty \leq \mu \leq \infty$
Binomial	$\mu(1-\mu)$	1	logit	$\log(\mu/(1-\mu))$	$0 \leq \mu \leq 1$
Poisson	μ	1	log	$\ln(\mu)$	$0 \leq \mu \leq \infty$
Gamma	μ^2	ν^{-1}	reciprocal	$1/\mu$	$-\infty \leq \mu \leq \infty$

The y_i are assumed to be independently and identically distributed so that the log likelihood function is simply the sum of contributions $l_i = \log f(y_i; \theta_i, \phi)$ from the individual observations

$$l(\boldsymbol{\theta}; \mathbf{y}) = \sum_i \left(\theta_i y_i - b(\theta_i)\right)/\left(a(\phi) + c(y_i, \phi)\right) \tag{7.5}$$

The mean and variance of Y can be derived from the relations

$$E\left(\frac{\partial l}{\partial \theta}\right) = 0 \tag{7.6}$$

and

$$E\left(\frac{\partial^2 l}{\partial \theta^2}\right) + E\left(\frac{\partial l}{\partial \theta}\right)^2 = 0 \tag{7.7}$$

and are given by

$$E(Y) = b'(\theta) = \mu \tag{7.8}$$

and

$$\mathrm{var}(Y) = b''(\theta)a(\phi) = V(\mu)a(\phi) \tag{7.9}$$

where $b'(\theta)$ and $b''(\theta)$ denote the fist and second derivative of $b(\theta)$ with respect to θ and the variance function $V(\mu)$ is obtained by expressing $b''(\theta)$ as a function of μ. It can be seen from equation (7.4) that the variance for the normal distribution is simply σ^2 regardless of the value of the mean μ, i.e., the variance function is 1.

The data on Australian school children will be analyzed by assuming a Poisson distribution for the number of days absent from school. The Poisson distribution represents the probability of observing y events if these events occur independently in continuous time at a constant instentaneous probability rate (or incidence rate); see, for example, Clayton and Hills (1993). The Poisson distribution is given by

$$f(y;\mu) = \mu^y e^{-\mu}/y!, \quad y = 0, 1, 2, \cdots. \tag{7.10}$$

Taking the logarithm and adding over observations, the log likelihood is

$$l(\boldsymbol{\mu}; \mathbf{y}) = \sum_i \left\{ (y_i \ln \mu_i - \mu_i) - \ln (y_i!) \right\} \tag{7.11}$$

so that $\theta = \ln \mu$, $b(\theta) = \exp(\theta)$, $\phi = 1$, $a(\phi) = 1$ and $\mathrm{var}(y) = \exp(\theta) = \mu$. Therefore, the variance of the Poisson distribution is not constant, but equal to the mean. Unlike the normal distribution, the Poisson distribution has no separate parameter for the variance and the same is true of the binomial distribution. Table 7.2 shows the variance functions and dispersion parameters for some commonly used probability distributions.

7.2.1 Model selection and measure of fit

Lack of fit can be expressed by the deviance, which is minus twice the difference between the maximized log likelihood of the model and the maximum likelihood achievable, i.e., the maximized likelihood of the *full* or saturated model. For the normal distribution, the deviance is simply the residual sum of squares. Another measure of lack of fit is the generalized Pearson X^2,

$$X^2 = \sum_i \left(y_i - \hat{\mu}_i \right)^2 / V (\hat{\mu}_i) \tag{7.12}$$

which, for the Poisson distribution, is just the familiar statistic for two-way cross-tabulations (see Chapter 2). Both the deviance and Pearson X^2 have χ^2 distributions when the sample size tends to infinity. When the dispersion parameter ϕ is fixed (not estimated), an analysis of deviance can be used for testing nested models in the same way as analysis of variance is used for linear models. The difference in deviance between two models is simply compared with the χ^2 distribution with degrees of freedom equal to the difference in model degrees of freedom.

The Pearson and deviance residuals are defined as the (signed) square roots of the contributions of the individual observations to the Pearson X^2 and deviance, respectively. These residuals can be used to assess the appropriateness of the link and variance functions.

A relatively common phenomenon with binary and count data is *overdispersion*, i.e., the variance is greater than that of the assumed distribution (binomial and Poisson, respectively). This overdispersion may be due to extra variability in the true mean parameter μ that has not been completely explained by the covariates. One way of addressing the problem is to assume a (prior) distribution of this parameter and to assume that conditional on the parameter having a certain value, the response variable follows the binomial (or Poisson) distribution. Such models are called *random effects models*.

A more pragmatic way of accommodating this overdispersion in the model is to assume that the variance is proportional to the variance function, but to estimate the dispersion rather than assuming the value 1 appropriate for the distributions. For the Poisson distribution, the variance is modeled as

$$\text{var}(Y) = \phi\mu \tag{7.13}$$

where ϕ is estimated from the Deviance or Pearson X^2. (This is analogous to the estimation of the residual variance in linear regresion models from the residual sums of squares.) This parameter $\hat{\phi}$ is then used to scale the estimated standard errors of the regression coefficients. If the variance is not proportional to the variance function, robust standard errors can be used (see next section). This approach of assuming a variance function that does not correspond to any probability distribution is an example of *quasi-likelihood* (see also Chapter 9).

For more details on general linear models, see McCullagh and Nelder (1989).

7.2.2 Robust standard errors of parameter estimates

A very useful feature of Stata is the ease with which robust standard errors can be obtained. We briefly outline two methods that have been implemented in Stata: the Huber-White sandwich estimator and bootstrapping.

The parameters β are estimated by maximizing the likelihood, i.e., by solving the set of equations

$$u(\hat{\beta}_k) = \frac{\partial l}{\partial \beta_k} \Big|_{\beta_k = \hat{\beta}_k} = 0, \quad k = 1, \cdots, p \qquad (7.14)$$

The quantities on the left-hand side are called *scores*. Equations (7.6) and (7.7) hold if the assumed likelihood is the true likelihood, and the same equations are then also satisfied if θ is replaced by β_k (apply the chain rule $\partial l / \partial \beta_k = \partial l / \partial \theta \times \partial \theta / \partial \beta_k$). It follows that, if the distributional assumptions are met, then

$$E(\mathbf{u}) = 0 \qquad (7.15)$$

and

$$\mathrm{var}(\mathbf{u})_{jk} = -E \left(\frac{\partial^2 l}{\partial \beta_j \partial \beta_k} \right) = -\mathbf{I}_{jk} \qquad (7.16)$$

where $\mathbf{u}$ is the score vector $(u(\hat{\beta}_1), \cdots, u(\hat{\beta}_p))^T$ and $-\mathbf{I}$ is minus the expectation of the Hessian of the log-likelihood function (the matrix of second derivatives), also known as the Fisher information.

The covariance matrix of the parameter estimates can be derived from the covariance matrix of the scores (using a first-order Taylor expansion, or the delta-method) as

$$\mathrm{var}(\boldsymbol{\beta}) = \mathbf{I}^{-1} \mathrm{var}(\mathbf{u}) \mathbf{I}^{-1} \qquad (7.17)$$

If equation (7.16) is true, the covariance matrix is simply $-\mathbf{I}^{-1}$ and these are the model-based (also known as "naive") standard errors for generalized linear models.

However, equation (7.16) is only true if the likelihood is the true likelihood of the data. If this assumption is not correct, due to misspecification of the co-variates, the link function, or the probability distribution function, we can still use equation (7.17) to obtain robust estimates of the standard errors. Instead of substituting equation (7.16) for the covariance matrix of the scores, we substitute the empirical covariance matrix derived from the observed contributions to the scores from each observation:

$$u_i(\hat{\beta}_k) = \frac{\partial l_i(y_i)}{\partial \beta_k} \Big|_{\beta_k = \hat{\beta}_k} \qquad (7.18)$$

This robust estimator of the standard errors is known as Huber, White or sandwich variance estimator, the latter name deriving from the form of equation (7.17) (for more details, see Binder, 1983).

In the description of the robust variance estimator in the *Stata User's Guide* (Section 26.10), it is pointed out that the use of robust standard errors implies a slightly less ambitious interpretation of the parameter estimates and their standard errors than a model-based approach. The parameter estimates are unbiased estimates of the estimates that would be obtained if we had an infinite sample (not any true parameters), and their standard errors are the standard deviations under repeated sampling followed by estimation (see also the FAQ by Sribney (1998)).

Another approach to estimating the standard errors without making any distributional assumptions is *bootstrapping* (Efron and Tibshirani, 1993). If we could obtain repeated samples from the population (from which our data was sampled), we could obtain an empirical sampling distribution of the parameter estimates. In Monte Carlo simulation, the required samples are drawn from the assumed distribution. In bootstrapping, the sample is resampled "to approximate what would happen if the population were sampled" (Manley, 1997). Bootstrapping works as follows. Take a random sample of n observations, with replacement, and estimate the regression coefficients. Repeat this a number of times to obtain a sample of estimates. From this sample, estimate the variance-covariance matrix of the parameter estimates. Confidence intervals can be constructed using the estimated variance or directly from the appropriate centiles of the empirical distribution of parameter estimates. See Manley (1997) and Efron and Tibshirani (1993) for more information on the bootstrap.

7.3 Analysis using Stata

7.3.1 Datasets from previous chapters

First, we show how linear regression can be carried out using glm. In Chapter 3, the U.S. air-pollution data were read in using the instructions

```
infile str10 town so2 temp manuf pop wind precip days
using usair.dat
drop if town=="Chicago"
```

and now we regress *so2* on a number of variables using

```
glm so2 temp pop wind precip, fam(gauss) link(id)
```

```
Iteration 1 : deviance = 10150.1520

Residual df   =        35                    No. of obs =          40
Pearson X2    =   10150.15                   Deviance   =    10150.15
Dispersion    =   290.0043                   Dispersion =    290.0043

Gaussian (normal) distribution, identity link
-----------------------------------------------------------------------------
     so2 |      Coef.   Std. Err.       t    P>|t|     [95% Conf. Interval]
---------+-------------------------------------------------------------------
    temp | -1.810123    .4404001    -4.110   0.000    -2.704183    -.9160635
     pop |  .0113089    .0074091     1.526   0.136    -.0037323     .0263501
    wind | -3.085284    2.096471    -1.472   0.150    -7.341347     1.170778
  precip |  .5660172    .2508601     2.256   0.030     .0567441     1.07529
   _cons |  131.3386    34.32034     3.827   0.001     61.66458     201.0126
-----------------------------------------------------------------------------
(Model is ordinary regression, use fit or regress instead)
```

The results are identical to those of the regression analysis as the remark at the bottom of the output explains. The dispersion parameter represents the residual variance given under Residual MS in the analysis of variance table of the regression analysis in Chapter 3. We can estimate robust standard errors using regress with the option robust (the robust estimator is not yet available for glm):

```
reg so2 temp pop wind precip, robust
```

```
Regression with robust standard errors          Number of obs =      40
                                                F(  4,    35) =    9.72
                                                Prob > F      =  0.0000
                                                R-squared     =  0.3446
                                                Root MSE      =   17.03

           |            Robust
      so2 |    Coef.   Std. Err.      t     P>|t|     [95% Conf. Interval]
----------+-----------------------------------------------------------------
     temp |  -1.810123   .3462819    -5.227   0.000    -2.513113   -1.107134
      pop |   .0113089   .0084062     1.345   0.187    -.0057566    .0283743
     wind |  -3.085284  1.792976     -1.721   0.094    -6.72522     .554651
   precip |   .5660172   .1919587     2.949   0.006     .1763203    .955714
    _cons |  131.3386   19.23291      6.829   0.000    92.29368    170.3835
```

giving slightly different standard errors indicating that the assumption of identically normally distributed residuals may not be entirely satisfied.

We now show how an analysis of variance model can be fitted using glm, using the slimming clinic example of Chapter 5. The data are read using

```
infile cond status resp using slim.dat
```

and the full, saturated model can be obtained using

```
xi: glm resp i.cond*i.status, fam(gauss) link(id)
```

```
i.cond                Icond_1-2    (naturally coded; Icond_1 omitted)
i.status              Istatu_1-2   (naturally coded; Istatu_1 omitted)
i.cond*i.status       IcXs_#-#     (coded as above)
Iteration 1 : deviance = 1078.8481
Residual df  =        30                       No. of obs  =       34
Pearson X2   =  1078.848                        Deviance    = 1078.848
Dispersion   =   35.9616                        Dispersion  =   35.9616
Gaussian (normal) distribution, identity link

     resp |    Coef.   Std. Err.      t     P>|t|     [95% Conf. Interval]
----------+-----------------------------------------------------------------
  Icond_2 |   .6780002  3.234433      0.210   0.835    -5.927593    7.283594
 Istatu_2 |  6.128834   3.19204       1.920   0.064    -.3901823   12.64785
 IcXs_2_2 | -.2655002   4.410437     -0.060   0.952    -9.272815    8.741815
    _cons |    -7.298   2.68185      -2.721   0.011    -12.77507   -1.820931

(Model is ordinary regression, use fit or regress instead)
```

This result is identical to that obtained using the command

```
xi: regress resp i.cond*i.status
```

(see Chapter 5, Exercises).

We can obtain the F-statistics for the interaction term by saving the deviance of the above model (residual sum of squares) in a local macro and refitting the model with the interaction removed:

```
local dev0=$S_E_dev
```

```
xi: glm resp i.cond i.status, fam(gauss) link(id)
```

```
i.cond              Icond_1-2   (naturally coded; Icond_1 omitted)
i.status            Istatu_1-2  (naturally coded; Istatu_1 omitted)
Iteration 1 : deviance = 1078.9784
Residual df  =      31                        No. of obs =         34
Pearson X2   =  1078.978                      Deviance   =   1078.978
Dispersion   =  34.80576                      Dispersion =   34.80576
Gaussian (normal) distribution, identity link
------------------------------------------------------------------------------
   resp |     Coef.   Std. Err.      t     P>|t|     [95% Conf. Interval]
--------+---------------------------------------------------------------------
Icond_2 |  .5352102   2.163277    0.247   0.806    -3.876821    4.947242
Istatu_2 |  5.989762   2.167029    2.764   0.010     1.570077    10.40945
  _cons |  -7.199832   2.094584   -3.437   0.002    -11.47176   -2.927901
------------------------------------------------------------------------------
(Model is ordinary regression, use fit or regress instead)
```

The increase in deviance caused by the removal of the interaction term represents the sum of squares of the interaction term after eliminating the main effects:

```
local dev1=$S_E_dev
local ddev='dev1'-'dev0'
disp 'ddev'
```

```
.13031826
```

and the F-statistic is simply the mean sum of squares of the interaction term after eliminating the main effects, divided by the residual mean square of the full model. The numerator and denominator degrees of freedom are 1 and 30, respectively, so that F and the associated p-value can be obtained as follows:

```
local f=('ddev'/1)/('dev0'/30)
disp 'f'
```

```
.00362382
```

```
disp fprob(1,30,'f')
```

.95239706

The general method for testing the difference in fit of two nested generalized linear models, using the difference in deviance, is not appropriate here because the dispersion parameter $\phi = \sigma^2$ was estimated.

The logistic regression analysis of Chapter 6 can also be repeated using `glm`. First read the data as before, without replicating records.

```
infile fr1 fr2 fr3 fr4 using tumour.dat, clear
gen therapy=int((_n-1)/2)
sort therapy
by therapy: gen sex=_n
reshape long fr, i(therapy sex) j(outc)
gen improve=outc
recode improve 1/2=0 3/4=1
list
```

	therapy	sex	outc	fr	improve
1.	0	1	1	28	0
2.	0	1	2	45	0
3.	0	1	3	29	1
4.	0	1	4	26	1
5.	0	2	1	4	0
6.	0	2	2	12	0
7.	0	2	3	5	1
8.	0	2	4	2	1
9.	1	1	1	41	0
10.	1	1	2	44	0
11.	1	1	3	20	1
12.	1	1	4	20	1
13.	1	2	1	12	0
14.	1	2	2	7	0
15.	1	2	3	3	1
16.	1	2	4	1	1

The `glm` command can be used with weights just like other estimation commands, so that we can analyze the table using

```
glm improve therapy sex [fweight=fr], fam(binomial) link(logit)
```

```
Iteration 1 : deviance =   381.8102
Iteration 2 : deviance =   381.2637
Iteration 3 : deviance =   381.2634
Iteration 4 : deviance =   381.2634

Residual df  =        296                      No. of obs =        299
Pearson X2   =   298.7045                      Deviance   =   381.2634
Dispersion   =   1.009137                      Dispersion =   1.288052

Bernoulli distribution, logit link
-----------------------------------------------------------------------
improve |     Coef.   Std. Err.      z     P>|z|     [95% Conf. Interval]
--------+--------------------------------------------------------------
therapy | -.5022014   .2456898    -2.044   0.041     -.9837444  -.0206583
    sex | -.6543125   .3714737    -1.761   0.078     -1.382388   .0737625
  _cons |  .3858095    .451417     0.855   0.393     -.4989515   1.270571
-----------------------------------------------------------------------
```

The likelihood ratio test for *sex* can be obtained as follows:

```
local dev0=$S_E_dev
quietly glm improve therapy [fweight=fr], fam(binomial)
link(logit)
local dev1=$S_E_dev
dis 'dev1'-'dev0'
```

```
3.3459816
```

```
dis chiprob(1,'dev1'-'dev0')
```

```
.0673693
```

which gives the same result as in Chapter 6.

7.3.2 Australian school children

We can now analyze the data in Table 7.1. The data are available as a Stata file *quine.dta* and can therefore be read simply by using the command

```
use quine, clear
```

The variables are of type string and can be converted to numeric using the encode command as follows:

```
encode eth, gen(ethnic)
drop eth
encode sex, gen(gender)
drop sex
encode age, gen(class)
drop age
encode lrn, gen(slow)
drop lrn
```

The number of children in each of the combinations of categories of *gender*, *class*, and *slow* can be found using

```
table slow class ethnic, contents(freq) by(gender)
```

```
----------+-------------------------------------------------------
          |                      ethnic and class
gender    | ---------- A ----------      ---------- N ----------
and slow  |  F0    F1    F2    F3        F0    F1    F2    F3
----------+-------------------------------------------------------
F         |
      AL  |   4     5     1     9         4     6     1    10
      SL  |   1    10     8               1    11     9
----------+-------------------------------------------------------
M         |
      AL  |   5     2     7     7         6     2     7     7
      SL  |   3     3     4               3     7     3
----------+-------------------------------------------------------
```

This reveals that there were no "slow learners" in class **F3**. A table for the means and standard deviations is obtained using

```
table slow class ethnic, contents(mean days sd days) by(gender)
   format(%4.1f)
```

```
----------+-------------------------------------------------------
          |                      ethnic and class
gender    | ---------- A ----------      ---------- N ----------
and slow  |  F0    F1    F2    F3        F0    F1    F2    F3
----------+-------------------------------------------------------
F         |
      AL  | 21.2  11.4   2.0  14.6      18.5  11.0   1.0  13.5
          | 17.7   6.5        14.9      10.7   8.9        11.5
          |
      SL  |  3.0  22.6  36.4            25.0   6.0   6.2
          |       18.7  26.5                   4.2   5.0
----------+-------------------------------------------------------
M         |
      AL  | 13.0  10.5  27.4  27.1       5.3   3.5   9.1  27.3
          |  8.0   4.9  14.7  10.4       5.4   0.7   9.5  22.9
          |
      SL  |  9.0   9.0  37.0            30.0   6.1  29.3
          |  6.2   5.2  23.4            32.5   6.1   7.0
----------+-------------------------------------------------------
```

This table suggests that the variance associated with the Poisson distribution is not appropriate here since squaring the standard deviations (to get the variances) results in values that are greater than the means, i.e., overdispersion. In this case, the overdispersion is probably due to the fact that there is a substantial variability in children's tendency to miss days of school that cannot be explained by the variables included in the model.

Ignoring the problem of overdispersion, a generalized linear model with a Poisson family and a log link can be fitted using

```
glm days slow class ethnic gender, fam(pois) link(log)
```

```
Iteration 1 : deviance = 1796.1923
Iteration 2 : deviance = 1768.7236
Iteration 3 : deviance = 1768.6453
Iteration 4 : deviance = 1768.6453
Residual df   =       141                No. of obs =        146
Pearson X2    =  1990.103                Deviance   =   1768.645
Dispersion    =   14.1142                Dispersion =   12.54358
Poisson distribution, log link
-----------------------------------------------------------------------
    days |     Coef.   Std. Err.      z     P>|z|     [95% Conf. Interval]
---------+-------------------------------------------------------------
    slow |   .2661578   .0445711    5.972   0.000     .1788001    .3535155
   class |   .2094662   .0218242    9.598   0.000     .1666916    .2522408
  ethnic |  -.5511688   .0418388  -13.174   0.000    -.6331713   -.4691663
  gender |   .2256243   .0415924    5.425   0.000     .1441048    .3071438
   _cons |   2.336676   .1427907   16.364   0.000     2.056812    2.616541
-----------------------------------------------------------------------
```

The algorithm takes four iterations to converge to the maximum likelihood (or minimum deviance) solution. When the iteration history is not of interest, the option **nolog** can be used to stop it being given. The confidence intervals are likely to be too narrow because of overdispersion. McCullagh and Nelder (1989) use the Pearson X^2 divided by the degrees of freedom to estimate the dispersion parameter for Poisson models. This can be achieved using the option **scale(x2)**:

```
glm days slow class ethnic gender, fam(pois) link(log) scale(x2)
    nolog
```

```
Residual df   =       141                No. of obs =        146
Pearson X2    =  1990.103                Deviance   =   1768.645
Dispersion    =   14.1142                Dispersion =   12.54358

Poisson distribution, log link
-----------------------------------------------------------------------
    days |     Coef.   Std. Err.      z     P>|z|     [95% Conf. Interval]
---------+-------------------------------------------------------------
    slow |   .2661578   .1674485    1.589   0.112    -.0620352    .5943508
   class |   .2094662   .0819911    2.555   0.011     .0487667    .3701657
  ethnic |  -.5511688   .1571835   -3.507   0.000    -.8592428   -.2430948
  gender |   .2256243   .1562578    1.444   0.149    -.0806353    .531884
   _cons |   2.336676   .5364486    4.356   0.000     1.285256    3.388096
-----------------------------------------------------------------------
(Standard errors scaled using square root of Pearson X2-based dispersion)
```

Allowing for overdispersion has had no effect on the regression coefficients, but a large effect on the p-values and confidence intervals so that *gender* and *slow* are now no longer significant. These terms will be removed from the model. The coefficients can be interpreted as the difference in the logs of the predicted mean counts between groups. For example, the log of the predicted mean number of days absent from school for white children is -0.55 lower than that for Aboriginals.

$$\ln(\hat{\mu}_2) = \ln(\hat{\mu}_1) - 0.55 \qquad (7.19)$$

It is difficult to think in terms of the log of the mean. Instead of calculating the exponentials of the coefficients and their confidence limits, we can ask Stata to do so using the option eform as follows:

```
glm days class ethnic, fam(pois) link(log) scale(x2) eform nolog
```

```
Residual df  =     143              No. of obs =       146
Pearson X2   =  2091.267            Deviance    =  1823.481
Dispersion   =  14.62424            Dispersion =  12.75162


Poisson distribution, log link
-----------------------------------------------------------------------
   days |      IRR   Std. Err.      z    P>|z|     [95% Conf. Interval]
--------+--------------------------------------------------------------
  class | 1.177895   .0895243    2.154   0.031     1.014874    1.367102
  ethnic |  .5782531   .0924968   -3.424   0.001     .4226303     .79118
-----------------------------------------------------------------------
(Standard errors scaled using square root of Pearson X2-based dispersion)
```

Therefore, white children are absent from school about 58% as often as Aboriginal children (95% confidence interval from 42% to 79%) after controlling for *class*. We have treated *class* as a continuous measure. To see whether this appears to be appropriate, we can form the square of *class* and include this in the model:

```
gen class2=class^2
glm days class class2 ethnic, fam(pois) link(log) scale(x2)
eform nolog
```

```
Residual df  =     142              No. of obs =       146
Pearson X2   =  2081.259            Deviance    =   1822.56
Dispersion   =  14.65676            Dispersion =  12.83493


Poisson distribution, log link
-----------------------------------------------------------------------
   days |      IRR   Std. Err.      z    P>|z|     [95% Conf. Interval]
--------+--------------------------------------------------------------
  class | 1.059399  0.4543011   0.135   0.893     0.4571295    2.455158
 class2 | 1.020512  0.0825501   0.251   0.802     0.8708906    1.195839
 ethnic | 0.5784944  0.092643   -3.418   0.001     0.4226525   0.7917989
-----------------------------------------------------------------------
(Standard errors scaled using square root of Pearson X2-based dispersion)
```

This term is not significant, so we can return to the simpler model. (Note that the interaction between *class* and *ethnic* is also not significant, see exercises.)

We now look at the residuals for this model. Stata has a post-estimation function called glmpred that can be used to calculate deviance or Pearson residuals. The Pearson residual can be standardized by dividing by the square root of the estimated dispersion parameter that is stored in the macro $S_E_dc:

```
glmpred resp, pearson
gen stres=resp/sqrt($S_E_dc)
```

and plotted against the linear predictor using

```
glmpred xb, xb
graph stres xb, xlab ylab l1("Standardised Residuals") gap(3)
```

with the result shown in Figure 7.1.

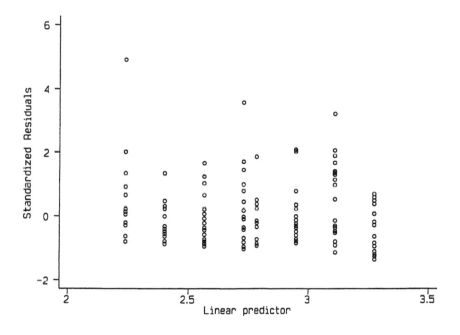

Figure 7.1 *Standardized residuals against linear predictor.*

There is one large outlier. In order to find out which observation this is, we list a number of variables for cases with large standardized Pearson residuals:

```
glmpred mu, mu
list stres days mu ethnic class if stres>2|stres<-2
```

	stres	days	mu	ethnic	class
45.	2.03095	53	19.07713	A	F1
46.	2.09082	54	19.07713	A	F1
58.	2.070247	60	22.47085	A	F2
59.	3.228685	81	22.47085	A	F2
72.	4.924755	67	9.365361	N	F0
104.	3.588988	69	15.30538	N	F3
109.	2.019528	33	9.365361	N	F0

Case 72, a white primary school child, has a very large residual.

We could check the assumptions of the model by estimating robust standard errors instead of assuming a variance function. Unfortunately, the robust sandwich estimator is presently not available for `glm` or for `poisson` (the custommade command for running Poisson regression). We therefore use bootstrap estimates. This can be done using the bootstrap command `bs`, followed by the estimation command in quotes, followed, in quotes, by expressions for the estimates for which bootstrap standard errors are required. To be on the safe side, ask for 500 bootstrap samples using the option `reps(500)`:

```
bs "poisson days class ethnic" "_b[class] _b[ethnic]", reps(500)
```

```
command:     poisson days class ethnic
statistics:  _b[class] _b[ethnic]
(obs=146)

Bootstrap statistics

Variable |  Reps   Observed     Bias   Std. Err.   [95% Conf. Interval]
---------+-------------------------------------------------------------
     bs1 |   500   .1637288  -.0055554  .0763474   .0137268   .3137308
(N)      |
         |                                         .0166459   .3092798
(P)      |
         |                                         .0272332   .3281089
(BC)     |
---------+-------------------------------------------------------------
     bs2 |   500  -.5477436  -.0016573  .1589607  -.8600583  -.2354289
(N)      |
         |                                        -.8715353  -.2338316
(P)      |
         |                                        -.8715353  -.2338316
(BC)     |
-----------------------------------------------------------------------
                 N = normal, P = percentile, BC = bias
corrected
```

This compares very well with the overdispersed Poisson model result:

```
glm days class ethnic, fam(pois) link(log) scale(x2) nolog
```

```
Residual df  =        143                    No. of obs =        146
Pearson X2   =  2091.267                     Deviance   =  1823.481
Dispersion   =  14.62424                     Dispersion =  12.75162

Poisson distribution, log link
--------------------------------------------------------------------------
    days |    Coef.   Std. Err.      z    P>|z|     [95% Conf. Interval]
---------+----------------------------------------------------------------
   class |  0.1637288  0.0760037   2.154   0.031    0.0147643    0.3126932
  ethnic | -0.5477436  0.159959   -3.424   0.001   -0.8612575   -0.2342298
   _cons |  3.168776   0.3170113   9.996   0.000    2.547446     3.790107
--------------------------------------------------------------------------
(Standard errors scaled using square root of Pearson X2-based dispersion)
```

We could also model overdispersion by assuming a *random effects* model where each child has an unobserved, random proneness to be absent from school. This proneness (called "frailty" in a medical context) is added to the linear predictor, causing some children to have higher or lower predicted rates of absence from school than those predicted by the explanatory variables. The observed counts are assumed to have Poisson distribution conditional on the random effects. If the frailties are assumed to have a gamma distribution, then the (marginal) distribution of the counts has a negative binomial distribution. The negative binomial model can be fitted using nbreg as follows:

```
nbreg days class ethnic, nolog
```

```
Negative Binomial Regression                 Number of obs   =        146
                                             Model chi2(2)   =      15.77
                                             Prob > chi2     =     0.0004
Log Likelihood =   -551.2462527              Pseudo R2       =     0.0141

--------------------------------------------------------------------------
    days |    Coef.   Std. Err.      z    P>|z|     [95% Conf. Interval]
---------+----------------------------------------------------------------
_lnmean  |
   class |  .1505165   .0732832    2.054   0.040    .0068841     .2941489
  ethnic | -.5414185   .1578378   -3.430   0.001   -.8507748    -.2320622
   _cons |  3.19392    .3217681    9.926   0.000    2.563266     3.824574
---------+----------------------------------------------------------------
_lnalpha |
   _cons | -.1759664   .1243878   -1.415   0.157   -.4197619     .0678292
--------------------------------------------------------------------------
   alpha   0.8386462   [ lnalpha]_cons = ln(alpha)
                 (LR test against Poisson, chi2(1) =  1309.466 P =
0.0000)
```

All three methods of analyzing the data lead to the same conclusions. The Poisson model is a special case of the negative binomial model with $\alpha = 0$. The likelihood ratio test for α is therefore a test of the negative binomial against the Poisson distribution. The very small p-value "against Poisson" indicates that there is significant overdispersion.

7.4 Exercises

1. Calculate the F-statistic and difference in deviance for *status* controlling for *cond* for the data in *slim.dat*.

2. Fit the model using only *status* as the independent variable, using robust standard errors. How does this compare with a t-test with unequal variances?

3. Test the significance of the interaction between *class* and *ethnic* for the data in *quine.dat*.

4. Excluding the outlier (case 72), fit the model with explanatory variables *ethnic* and *class*.

5. Find robust standard errors for the Poisson regression discussed in this chapter. If this option is not yet available in Stata, find a program on the internet; for example, `rglm` by Roger Newson.

6. Dichotomize days absent from school by classifying 14 days or more as frequently absent. Analyze this new response using both the logistic and probit link and the binomial family.

7. Use `logit` and `probit` to estimate the same models with robust standard errors and compare this with the standard errors obtained using bootstrapping.

See also the exercises in Chapter 10.

CHAPTER 8

Analysis of Longitudinal Data I: The Treatment of Postnatal Depression

8.1 Description of data

The dataset to be analyzed in this chapter originates from a clinical trial of the use of estrogen patches in the treatment of postnatal depression; full details are given in Gregoire et al. (1996). Sixty-one women with major depression, which began within 3 months of childbirth and persisted for up to 18 months postnatally, were allocated randomly to the active treatment or a placebo (a dummy patch); 34 received the former and the remaining 27 the latter. The women were assessed pretreatment and monthly for 6 months after treatment on the Edinburgh postnatal depression scale (EPDS), higher values of which indicate increasingly severe depression. The data are shown in Table 8.1; a value of −9 in this table indicates that the observation is missing.

Table 8.1 Data in *depress.dat*

subj	group	pre	dep1	dep2	dep3	dep4	dep5	dep6
1	0	18	17	18	15	17	14	15
2	0	27	26	23	18	17	12	10
3	0	16	17	14	−9	−9	−9	−9
4	0	17	14	23	17	13	12	12
5	0	15	12	10	8	4	5	5
6	0	20	19	11.54	9	8	6.82	5.05
7	0	16	13	13	9	7	8	7
8	0	28	26	27	−9	−9	−9	−9
9	0	28	26	24	19	13.94	11	9
10	0	25	9	12	15	12	13	20
11	0	24	14	−9	−9	−9	−9	−9
12	0	16	19	13	14	23	15	11
13	0	26	13	22	−9	−9	−9	−9
14	0	21	7	13	−9	−9	−9	−9
15	0	21	18	−9	−9	−9	−9	−9
16	0	22	18	−9	−9	−9	−9	−9
17	0	26	19	13	22	12	18	13
18	0	19	19	7	8	2	5	6

Table 8.1 Data in *depress.dat*

19	0	22	20	15	20	17	15	13.73
20	0	16	7	8	12	10	10	12
21	0	21	19	18	16	13	16	15
22	0	20	16	21	17	21	16	18
23	0	17	15	−9	−9	−9	−9	−9
24	0	22	20	21	17	14	14	10
25	0	19	16	19	−9	−9	−9	−9
26	0	21	7	4	4.19	4.73	3.03	3.45
27	0	18	19	−9	−9	−9	−9	−9
28	1	21	13	12	9	9	13	6
29	1	27	8	17	15	7	5	7
30	1	15	8	12.27	10	10	6	5.96
31	1	24	14	14	13	12	18	15
32	1	15	15	16	11	14	12	8
33	1	17	9	5	3	6	0	2
34	1	20	7	7	7	12	9	6
35	1	18	8	1	1	2	0	1
36	1	28	11	7	3	2	2	2
37	1	21	7	8	6	6.5	4.64	4.97
38	1	18	8	6	4	11	7	6
39	1	27.46	22	27	24	22	24	23
40	1	19	14	12	15	12	9	6
41	1	20	13	10	7	9	11	11
42	1	16	17	26	−9	−9	−9	−9
43	1	21	19	9	9	12	5	7
44	1	23	11	7	5	8	2	3
45	1	23	16	13	−9	−9	−9	−9
46	1	24	16	15	11	11	11	11
47	1	25	20	18	16	9	10	6
48	1	22	15	17.57	12	9	8	6.5
49	1	20	7	2	1	0	0	2
50	1	20	12.13	8	6	3	2	3
51	1	25	15	24	18	15.19	13	12.32
52	1	18	17	6	2	2	0	1
53	1	26	1	18	10	13	12	10
54	1	20	27	13	9	8	4	5
55	1	17	20	10	8.89	8.49	7.02	6.79
56	1	22	12	−9	−9	−9	−9	−9
57	1	22	15.38	2	4	6	3	3
58	1	23	11	9	10	8	7	4
59	1	17	15	−9	−9	−9	−9	−9

Table 8.1 Data in *depress.dat*

60	1	22	7	12	15	−9	−9	−9
61	1	26	24	−9	−9	−9	−9	−9

8.2 The analysis of longitudinal data

The data in Table 8.1 consist of repeated observations over time on each of
the 61 patients; such data are generally referred to as *longitudinal*. There is a
large body of methods that can be used to analyze longitudinal data, ranging
from the simple to the complex. Some useful references are Diggle et al. (1994),
Everitt (1995), and Hand and Crowder (1996). This chapter concentrates on
the following approaches:

- Graphical displays

- Summary measure or response feature analysis

In the next chapter, more formal modeling techniques will be applied to the
data.

8.3 Analysis using Stata

Assuming the data are in an ASCII file, *depress.dat* as listed in Table 8.1, they
may be read into Stata for analysis using the following instructions:

```
infile subj group pre dep1 dep2 dep3 dep4 dep5 dep6
using depress.dat
mvdecode _all, mv(-9)
```

The second of these instructions identifies the '−9's in the data as missing
values.

It is useful to begin examination of these data using the **summarize** procedure
to calculate means, variances, etc., within each of the two treatment groups;

```
summarize pre-dep6 if group==0
```

Variable	Obs	Mean	Std. Dev.	Min	Max
pre	27	20.77778	3.954874	15	28
dep1	27	16.48148	5.279644	7	26
dep2	22	15.88818	6.124177	4	27
dep3	17	14.12882	4.974648	4.19	22
dep4	17	12.27471	5.848791	2	23
dep5	17	11.40294	4.438702	3.03	18
dep6	17	10.89588	4.68157	3.45	20

```
summarize pre-dep6 if group==1
```

Variable	Obs	Mean	Std. Dev.	Min	Max
pre	34	21.24882	3.574432	15	28
dep1	34	13.36794	5.556373	1	27
dep2	31	11.73677	6.575079	1	27
dep3	29	9.134138	5.475564	1	24
dep4	28	8.827857	4.666653	0	22
dep5	28	7.309286	5.740988	0	24
dep6	28	6.590714	4.730158	1	23

There is a general decline in the EPDS over time in both groups, with the values in the active treatment group appearing to be consistently lower.

8.3.1 Graphical displays

A useful preliminary step in the analysis of longitudinal data is to graph the observations in some way. The aim is to highlight two particular aspects of the data, namely, how they evolve over time and how the measurements made at different times are related. A number of graphical displays can be used, including

- Separate plots of each subject's responses against time, differentiating in some way between subjects in different groups
- Boxplots of the observations at each time point by treatment group
- A plot of means and standard errors by treatment group for every time point
- A scatter-plot matrix of the repeated measurements

To begin, obtain the required scatter-plot matrix, identifying treatment groups with the labels 0 and 1, using

```
graph pre-dep6, matrix symbol([group]) ps(150)
```

The resulting plot is shown in Figure 8.1. The most obvious feature of this diagram is the increasingly strong relationship between the measurements of depression as the time interval between them decreases. This has important implications for the models appropriate for longitudinal data, as seen in Chapter 9.

In order to obtain the other graphs mentioned above, the data needs to be restructured from its present wide form to the long form, as described in the introductory chapter. This can be achieved very simply using the reshape command:

```
reshape long dep, i(subj) j(visit)
list in 1/13
```

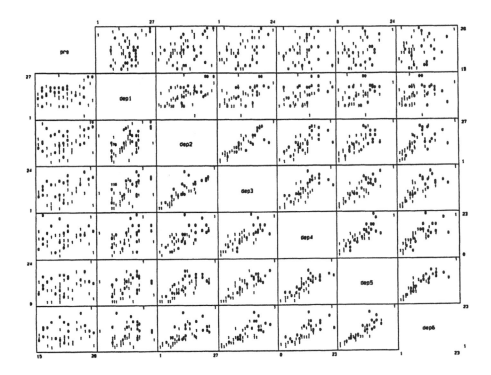

Figure 8.1 *Scatter-plot matrix for depression scores at six visits.*

	subj	visit	group	pre	dep
1.	1	1	0	18	17
2.	1	2	0	18	18
3.	1	3	0	18	15
4.	1	4	0	18	17
5.	1	5	0	18	14
6.	1	6	0	18	15
7.	2	1	0	27	26
8.	2	2	0	27	23
9.	2	3	0	27	18
10.	2	4	0	27	17
11.	2	5	0	27	12
12.	2	6	0	27	10
13.	3	1	0	16	17

There are a number of methods for obtaining diagrams containing separate plots of each individual's responses against time from the restructured data, but the simplest is to use the connect(L) command which 'tricks' Stata into drawing lines connecting points within a specified grouping variable in a single graph. Here, the required grouping variable is *subj*, the *y* variable is *dep*, and the *x* variable is *visit*. Before plotting, the data need to be sorted by the grouping variable and by the *x* variable:

```
sort group visit
graph dep visit, by(group) c(L)
```

The 'L' option connects points only so long as *visit* is ascending. For observations with *subj* equal to one, this is true; but for the second subject, *visit* begins at one again, so the last point for subject one is *not* connected with the first point for subject two. The remaining points for this subject are, however, connected and so on. Using the by(group) option with only two groups produces a display that is half empty; we therefore use the commands

```
sort visit
graph dep visit if group==0, c(L) ylab /*
       */ t1("placebo group") l1("depression") gap(3)
graph dep visit if group==1, c(L) ylab /*
       */ t1("estrogen patch group") l1("depression") gap(3)
```

to obtain the diagrams shown in Figure 8.2. The individual plots reflect the general decline in the depression scores over time indicated by the means obtained using the summarize command; there is, however, considerable variability. Notice that some plots are not complete because of the presence of missing values.

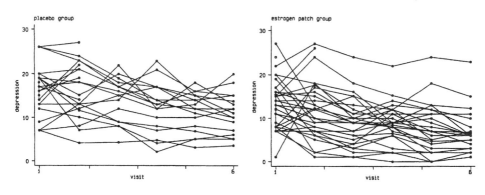

Figure 8.2 *Individual response profiles by treatment group.*

To obtain the boxplots of the depression scores at each visit for each treatment group, the following instructions can be used.

```
sort visit
graph dep if group==0, box by(visit) /*
       */ ylab t1("placebo group") t2(" ")
       l1("depression") gap(3)
graph dep if group==1, box by(visit) /*
       */ ylab t1("estrogen patch group") t2(" ")
       l1("depression") gap(3)
```

The resulting graphs are shown in Figure 8.3. Again, the general decline in depression scores in both treatment groups can be seen and, in the active treatment group, there is some evidence of outliers that may need to be examined. (Figure 8.2 shows that four of the outliers are due to one subject whose response profile lies above the others.)

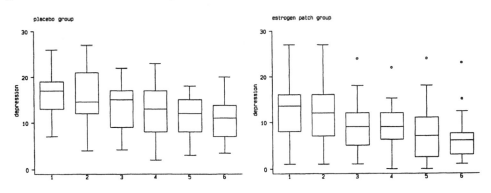

Figure 8.3 *Boxplots for six visits by treatment group.*

To obtain a plot of the mean profiles of each treatment group, which includes information about the standard errors of each mean, requires the use of the collapse instruction that produces a dataset consisting of selected summary statistics. Here, we need the mean depression score on each visit for each group, the corresponding standard deviations, and a count of the number of observations on which these two statistics are based. The necessary command is

```
collapse (mean) dep (sd) sddep=dep (count) n=dep, by(visit group)
list in 1/10
```

	visit	group	dep	sddep	n
1.	1	0	16.48148	5.279644	27
2.	1	1	13.36794	5.556373	34
3.	2	0	15.88818	6.124177	22
4.	2	1	11.73677	6.575079	31
5.	3	0	14.12882	4.974648	17
6.	3	1	9.134138	5.475564	29
7.	4	0	12.27471	5.848791	17
8.	4	1	8.827857	4.666653	28
9.	5	0	11.40294	4.438702	17
10.	5	1	7.309286	5.740988	28

The mean value is now stored in *dep*; but since more than one summary statistic for the depression scores was required, the remaining statistics were given new names in the collapse instruction.

The required mean and standard error plots can now be found using

```
sort group
gen high=dep+2*sddep/sqrt(n)
gen low=dep-2*sddep/sqrt(n)
graph dep high low visit, by(group) c(lll) sort
```

Again, we can obtain better-looking graphs by plotting them separately by group.

```
graph dep high low visit if group==0, c(lll) sort s(oii)
       pen(322) /* */ yscale(5,20) ylab ll("depression")
       gap(3) t1("placebo group")
graph dep high low visit if group==1, c(lll) sort s(oii)
       pen(322) /* */ yscale(5,20) ylab ll("depression")
       gap(3) t1("estrogen patch
group")
```

Here, we have set the range of the y-axes to the same values using the option yscale and have used the option pen() to display the confidence limits in the same color. Before printing, we have changed the line thickness of pen 3 to 4 units (click into **Prefs** in the menu bar, select "Graph Preferences," etc.). The resulting diagrams are shown in Figure 8.4.

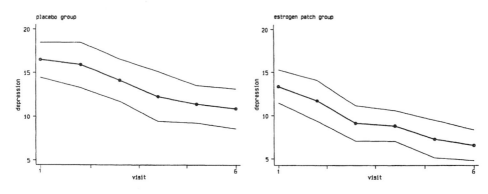

Figure 8.4 *Mean and standard error plots; the envelopes represent ± 2 standard errors.*

8.3.2 Response feature analysis

A relatively straightforward approach to the analysis of longitudinal data is that involving the use of *summary measures*, sometimes known as *response feature analysis*. The responses of each subject are used to construct a single number that characterizes some relevant aspect of the subject's response profile. (In some situations, more than a single summary measure may be required.) The summary measure needs to be chosen prior to the analysis of the data. The most commonly used measure is the mean of the responses over time since many investigations, e.g., clinical trials, are most concerned with differences in overall levels rather than more subtle effects. Other possible summary measures are listed in Mathews et al. (1990) and are shown here in Table 8.2.

Having identified a suitable summary measure, the analysis of the data generally involves the application of a simple univariate test (usually a t-test or

Table 8.2 Response features suggested in Mathews et al. (1990)

Type of data	Property to be compared between groups	Summary measure
Peaked	Overall value of response	Mean or area under curve
Peaked	Value of most extreme response	Maximum (minimum)
Peaked	Delay in response	Time to maximum or minimum
Growth	Rate of change of response	Linear regression coefficient
Growth	Final level of response	Final value or (relative) difference between first and last
Growth	Delay in response	Time to reach a particular value

its nonparametric equivalent) for group differences on the single measure now available for each subject. For the estrogen patch trial data, the mean over time seems an obvious summary measure. The mean of all non-missing values is obtained using

```
egen avg=rmean(dep1 dep2 dep3 dep4 dep5 dep6)
```

The differences between these means can be tested using a t-test assuming equal variances in the populations:

```
ttest avg, by(group)
```

```
Two-sample t test with equal variances

-----------------------------------------------------------------------------
   Group |     Obs        Mean    Std. Err.   Std. Dev.   [95% Conf. Interval]
---------+-------------------------------------------------------------------
       0 |      27    14.75605    .8782852    4.563704    12.95071    16.56139
       1 |      34    10.55206    .9187872    5.357404    8.682772    12.42135
---------+-------------------------------------------------------------------
combined |      61    12.41284    .6923949    5.407777    11.02785    13.79784
---------+-------------------------------------------------------------------
    diff |             4.20399    1.294842                1.613017    6.794964
-----------------------------------------------------------------------------
Degrees of freedom: 59

                   Ho: mean(0) - mean(1) = diff = 0

    Ha: diff < 0              Ha: diff ~= 0              Ha: diff > 0
      t =   3.2467              t =   3.2467               t =   3.2467
  P < t =   0.9990          P > |t| =   0.0019          P > t =   0.0010
```

Now relax the assumption of equal variances:

```
ttest avg, by(group) unequal
```

```
Two-sample t test with unequal variances

-----------------------------------------------------------------------
  Group |    Obs        Mean    Std. Err.    Std. Dev.   [95% Conf. Interval]
--------+--------------------------------------------------------------
      0 |     27    14.75605     .8782852     4.563704    12.95071    16.56139
      1 |     34    10.55206     .9187872     5.357404    8.682772    12.42135
--------+--------------------------------------------------------------
combined|     61    12.41284     .6923949     5.407777    11.02785    13.79784
--------+--------------------------------------------------------------
   diff |               4.20399    1.271045                1.660343    6.747637
-----------------------------------------------------------------------
Satterthwaite's degrees of freedom:  58.6777

                    Ho: mean(0) - mean(1) = diff = 0

   Ha: diff < 0                Ha: diff ~= 0               Ha: diff > 0
     t =    3.3075               t =    3.3075               t =    3.3075
   P < t =  0.9992         P > |t| =   0.0016         P > t =    0.0008
```

Both tests and the associated confidence intervals indicate clearly that there is a substantial difference in overall level in the two treatment groups.

8.4 Exercises

1. How would you produce the corresponding boxplots to those shown in Figure 8.3 using the data in the wide form?

2. Compare the results of the t-tests given in the text with the corresponding t-tests calculated only for those subjects having observations on all six post-randomization visits.

3. Repeat the summary measures analysis described in the text using now the mean over time divided by the standard deviation over time.

4. Test for differences in the mean over time controlling for the baseline measurement using

   a. A change score defined as the difference between the mean over time and the baseline measurement;

   b. Analysis of covariance of the mean over time using the baseline measurement as a covariate.

See also exercises in Chapter 9.

CHAPTER 9

Analysis of Longitudinal Data II: Epileptic Seizures and Chemotherapy

9.1 Introduction

In a clinical trial reported by Thall and Vail (1990), 59 patients with epilepsy were randomized to groups receiving either the anti-epileptic drug progabide, or a placebo, as an adjutant to standard chemotherapy. The number of seizures was counted over four 2-week periods. In addition, a baseline seizure rate was recorded for each patient, based on the 8-week prerandomization seizure count. The age of each patient was also noted. The data are shown in Table 9.1. (These data also appear in Hand et al., 1994.)

Table 9.1 Data in *chemo.dat*

subj	y1	y2	y3	y4	treatment	baseline	age
1	5	3	3	3	0	11	31
2	3	5	3	3	0	11	30
3	2	4	0	5	0	6	25
4	4	4	1	4	0	8	36
5	7	18	9	21	0	66	22
6	5	2	8	7	0	27	29
7	6	4	0	2	0	12	31
8	40	20	23	12	0	52	42
9	5	6	6	5	0	23	37
10	14	13	6	0	0	10	28
11	26	12	6	22	0	52	36
12	12	6	8	4	0	33	24
13	4	4	6	2	0	18	23
14	7	9	12	14	0	42	36
15	16	24	10	9	0	87	26
16	11	0	0	5	0	50	26
17	0	0	3	3	0	18	28
18	37	29	28	29	0	111	31
19	3	5	2	5	0	18	32
20	3	0	6	7	0	20	21
21	3	4	3	4	0	12	29

Table 9.1 Data in *chemo.dat*

22	3	4	3	4	0	9	21
23	2	3	3	5	0	17	32
24	8	12	2	8	0	28	25
25	18	24	76	25	0	55	30
26	2	1	2	1	0	9	40
27	3	1	4	2	0	10	19
28	13	15	13	12	0	47	22
29	11	14	9	8	1	76	18
30	8	7	9	4	1	38	32
31	0	4	3	0	1	19	20
32	3	6	1	3	1	10	30
33	2	6	7	4	1	19	18
34	4	3	1	3	1	24	24
35	22	17	19	16	1	31	30
36	5	4	7	4	1	14	35
37	2	4	0	4	1	11	27
38	3	7	7	7	1	67	20
39	4	18	2	5	1	41	22
40	2	1	1	0	1	7	28
41	0	2	4	0	1	22	23
42	5	4	0	3	1	13	40
43	11	14	25	15	1	46	33
44	10	5	3	8	1	36	21
45	19	7	6	7	1	38	35
46	1	1	2	3	1	7	25
47	6	10	8	8	1	36	26
48	2	1	0	0	1	11	25
49	102	65	72	63	1	151	22
50	4	3	2	4	1	22	32
51	8	6	5	7	1	41	25
52	1	3	1	5	1	32	35
53	18	11	28	13	1	56	21
54	6	3	4	0	1	24	41
55	3	5	4	3	1	16	32
56	1	23	19	8	1	22	26
57	2	3	0	1	1	25	21
58	0	0	0	0	1	13	36
59	1	4	3	2	1	12	37

9.2 Possible models

The data listed in Table 9.1 consist of repeated observations on the same subject taken over time, and are a further example of a set of *longitudinal data*. During the last decade, statisticians have considerably enriched the methodology available for the analysis of such data (see Diggle, Liang, and Zeger (1994)); and many of these developments are implemented in Stata.

Models for the analysis of longitudinal data are similar to the generalized linear models encountered in Chapter 7, but with one very important difference, namely, the residual terms are allowed to be correlated rather than independent. This is necessary since the observations at different time points in a longitudinal study involve the same subjects, thus generating some pattern of dependence that needs to be accounted for by any proposed model.

9.2.1 Normally distributed responses

Suppose that a normally distributed response is observed on each individual at T time points, then the basic regression model for longitudinal data becomes (cf. equation (3.2))

$$\mathbf{y}_i = \mathbf{X}\boldsymbol{\beta} + \boldsymbol{\epsilon}_i \qquad (9.1)$$

where $\mathbf{y}_i^T = (y_{i1}, y_{i2}, \cdots, y_{iT})$, $\boldsymbol{\epsilon}^T = (\epsilon_{i1}, \cdots, \epsilon_{iT})$, $\mathbf{X}$ is a $T \times (p+1)$ design matrix and $\boldsymbol{\beta}^T = (\beta_0, \cdots, \beta_p)$ is a vector of parameters. Assuming the residual terms have a multivariate normal distribution with a particular covariance matrix allows maximum likelihood estimation to be used; details are given in Jennrich and Schluchter (1986). If all covariance parameters are estimated independently, giving an unstructured covariance matrix, then this approach is essentially equivalent to multivariate analysis of variance for longitudinal data.

Another approach is to try to explain the covariances by introducing latent variables, the simplest example being a *random intercept* model given by

$$y_{ij} = \boldsymbol{\beta}^T \mathbf{x}_{ij} + u_i + \epsilon_{ij} \qquad (9.2)$$

where the latent random effects u_i and the residuals ϵ_{ij} are assumed to be independently normally distributed with zero means and constant variances. This random effects model implies that the covariances of the responses y_{ij} and y_{ik} at different time points are all equal to each other and that the variances at each time point are constant, a structure of the covariance matrix known as *compound symmetry* (see, for example, Winer, 1971). We can specify compound symmetry or any other structure for the covariance matrix (which may not correspond to any a random effects model) and estimate the parameters by maximum likelihood. If compound symmetry is assumed, this is essentially equivalent to assuming a split-plot design and carrying out "averaged univariate analysis of variance."

9.2.2 Non-normal responses

In cases where normality cannot be assumed, it is not possible to specify a likelihood with an arbitrary correlation structure. We can define random effects models by introducing the random intercept into the linear predictor,

$$\eta_{ij} = \beta^T \mathbf{x}_{ij} + u_i \tag{9.3}$$

where the u_i are independently distributed. (The negative binomial model is an example of the model above where there is only one observation per subject, see Chapter 7.) The attraction of such models, also known as generalized linear mixed models, is that they correspond to a probabilistic mechanism that may have generated the data and that estimation is via maximum likelihood. However, generalized linear mixed models tend to be difficult to estimate (see, for example, Goldstein, 1995) and the implied correlation structures are not sufficiently general for all purposes.

In the generalized estimating equation approach introduced by Liang and Zeger (1986), any required covariance structure and any link function can be assumed and parameters estimated without specifying the joint distribution of the repeated observations. Estimation is via a quasi-likelihood approach (see Wedderburn, 1974). This is analogous to the approach taken in Chapter 7 where we freely estimated the dispersion ϕ for a Poisson model that is characterized by $\phi = 1$.

Since the parameters specifying the structure of the correlation matrix are rarely of great practical interest (they are what is known as *nuisance parameters*), simple structures are used for the within-subject correlations giving rise to the so-called *working correlation matrix*. Liang and Zeger show that the estimates of the parameters of most interest, i.e., those that determine the mean profiles over time, are still valid even when the correlation structure is incorrectly specified.

The two approaches—random effects modeling and generalized estimating equations—lead to different interpretations of between subject effects. In random effects models, a between-subject effect represents the difference between subjects conditional on having the same random effect, whereas the parameters of generalized estimating equations represent the average difference between subjects. The two model types are therefore also known as *conditional* and *marginal* models, respectively. In practice, this distinction is important only if link functions other than the identity or log link are used, for example, in logistic regression (see Diggle et al., 1994).

A further issue with many longitudinal datasets is the occurrence of dropouts, i.e., subjects who fail to complete all scheduled visits. (The depression data of the previous chapter suffered from this problem.) A taxonomy of dropouts is given in Diggle, Liang, and Zeger (1994) where it is shown that it is necessary to make particular assumptions about the dropout mechanism for the analyses described in this chapter to be valid.

9.3 Analysis using Stata

Random effects models have not yet been implemented in Stata, except for linear regression. We will therefore focus on the generalized estimating equation approach to modeling longitudinal data that has been implemented in the `xtgee` module of Stata. The main components of a model that have to be specified are

- The assumed distribution of the residual terms, specified in the `family` option.

- The link between the response variable and its linear predictor, specified in the `link` option.

- The structure of the working correlation matrix, specified in the `correlations` option.

In general, it is not necessary to specify both `family` and `link` since, as explained in Chapter 7, the default link is the canonical link for the specified family.

The `xtgee` module will most often be used with the `family(gauss)` option, together with the identity link function, giving rise to multivariate normal regression, the multivariate analog of multiple regression as described in Chapter 3. This option is illustrated on the postnatal depression data used in the previous chapter.

9.3.1 Postnatal depression data

The data are obtained using the instructions

```
infile subj group pre dep1 dep2 dep3 dep4 dep5 dep6
using depress.dat
mvdecode _all, mv(-9)
reshape long dep, i(subj) j(visit)
```

To begin, fit a model that regresses depression on *group, pre*, and *visit* under the unrealistic assumptions of independence. The necessary instruction written out in its fullest form is

```
xtgee dep group pre visit, i(subj) t(visit) corr(indep) /*
    */ link(iden) fam(gauss)
```

```
Iteration 1: tolerance = 1.174e-14

General estimating equation for panel data      Number of obs    =       295
Group variable:                          subj   Number of groups =        61
Link:                                identity   Obs/group, min   =         1
Family:                              Gaussian                avg   =      4.84
Correlation:                      independent                max   =         6
                                                chi2(3)          =    144.15
Scale parameter:                    25.80052    Prob > chi2      =    0.0000
Pearson chi2(291):                   7507.95    Deviance         =   7507.95
Dispersion (Pearson):               25.80052    Dispersion       =  25.80052


------------------------------------------------------------------------------
    dep |      Coef.   Std. Err.       z     P>|z|     [95% Conf. Interval]
--------+---------------------------------------------------------------------
  group |  -4.290664   .6072954    -7.065    0.000    -5.480941    -3.100387
    pre |   .4769071   .0798565     5.972    0.000     .3203913     .633423
  visit |  -1.307841    .169842    -7.700    0.000    -1.640725    -.9749569
  _cons |   8.233577   1.803945     4.564    0.000     4.697909     11.76924
------------------------------------------------------------------------------
```

Here, the fitted model is simply the multiple regression model described in Chapter 3 for 295 observations that are assumed to be independent of one another; the scale parameter is equal to the residual mean square and the deviance is equal to the residual sum of squares. The estimated regression coefficients and their associated standard errors indicate that the covariates *group, pre*, and *visit* are all significant. However, treating the observation as independent is unrealistic and will almost certainly lead to poor estimates of the standard errors. Standard errors for between-subject factors (here, *group* and *pre*) tend to be underestimated because we are treating observations from the same subject as independent, thus increasing the apparent sample size; standard errors for within subject factors (here, *visit*) tend to be overestimated.

Now abandon the assumption of independence and estimate a correlation matrix having compound symmetry (i.e., constrain the correlations between the observations at any pair of time points to be equal). Such a correlational structure is introduced using corr(exchangeable), or the abbreviated form corr(exc). The model can be fitted as follows:

```
xtgee dep group pre visit, i(subj) t(visit) corr(exc) /*
     */ link(iden) fam(gauss)
```

Instead of specifying the subject and time identifiers using the options i() and t(), we can also declare the data as being of the form xt (for cross-sectional time series) as follows:

```
iis subj
tis visit
```

Since both the link and family corespond to the default options, the same analysis can be carried out using the shorter command

```
xtgee dep group pre visit, corr(exc)
```

```
Iteration 1: tolerance = .0461802
Iteration 2: tolerance = .0003057
Iteration 3: tolerance = 2.409e-06
Iteration 4: tolerance = 1.900e-08

General estimating equation for panel data    Number of obs    =      295
Group variable:                        subj    Number of groups =       61
Link:                              identity    Obs/group, min   =        1
Family:                            Gaussian                avg   =     4.84
Correlation:                   exchangeable                max   =        6
                                               chi2(3)          =   124.86
Scale parameter:                    25.90007   Prob > chi2      =   0.0000
Pearson chi2(291):                   7536.92   Deviance         =  7536.92
Dispersion (Pearson):               25.90007   Dispersion       = 25.90007

-----------------------------------------------------------------------------
    dep |      Coef.   Std. Err.       z     P>|z|     [95% Conf. Interval]
--------+--------------------------------------------------------------------
  group |  -4.044785   1.054732    -3.835    0.000    -6.112023   -1.977548
    pre |   .4608146   .1404686     3.281    0.001     .1855012    .7361279
  visit |  -1.232721   .1249349    -9.867    0.000    -1.477588   -.9878527
  _cons |   8.424423    3.04519     2.766    0.006     2.455961   14.39289
-----------------------------------------------------------------------------
```

After estimation, xtcorr reports the working correlation matrix

```
xtcorr
```

```
Estimated within subj correlation matrix R:

            c1       c2       c3       c4       c5       c6
r1   1.0000
r2   0.5010   1.0000
r3   0.5010   0.5010   1.0000
r4   0.5010   0.5010   0.5010   1.0000
r5   0.5010   0.5010   0.5010   0.5010   1.0000
r6   0.5010   0.5010   0.5010   0.5010   0.5010   1.0000
```

Note that that the standard errors for *group* and *pre* have increased, whereas that for *visit* has decreased as expected. The estimated within-subjects correlation matrix demonstrates the compound symmetry structure.

As we pointed out before, the compound symmetry structure arises when a random subject effect is introduced into the linear model for the observations. Such a random effect model can be fitted using xtreg

```
xtreg dep group pre visit, i(subj)
```

```
                                          Random-effects GLS regression
sd(u_subj)                  =   3.777959       Number of obs =      295
sd(e_subj_t)                =   3.355772                   n =       61
sd(e_subj_t + u_subj)       =   5.053135               T-bar = 3.15517

corr(u_subj, X)             =   0 (assumed)       R-sq within   =   0.2984
                                                      between   =   0.3879
                                                      overall   =   0.3312
------------------- theta --------------------
   min     5%     median      95%      max       chi2(  3)      =   134.33
  0.3359  0.3359  0.6591     0.6591   0.6591      Prob > chi2 =   0.0000

------------------------------------------------------------------------------
     dep |     Coef.    Std. Err.       z     P>|z|     [95% Conf. Interval]
---------+--------------------------------------------------------------------
   group |  -4.023373   1.089009     -3.695   0.000     -6.157792   -1.888954
     pre |   .4598446   .1452149      3.167   0.002      .1752288    .7444605
   visit |  -1.22638    .1176654    -10.423   0.000     -1.457      -.9957599
   _cons |   8.433315   3.143717      2.683   0.007      2.271742   14.59489
------------------------------------------------------------------------------
```

giving almost the same parameter estimates as the GEE model. We also obtain the variance components, with the between-subject standard deviation, sd(u_subj), estimated as 3.78 and the within-subject standard deviation, sd(e_subj_t), estimated as 3.36. The correlation between time-points predicted by this model is the ratio of the between-subject variance to the total variance and is calculated using

    disp 3.777959^2/ 5.053135^2

    .55897538

The compound symmetry structure for the correlations implied by the random effects model is frequently not acceptable since it is more likely that correlations between pairs of observations widely separated in time will be lower than for observations made closer together. This pattern was apparent from the scatter-plot matrix given in the previous chapter.

To allow for a more complex pattern of correlations among the repeated observations, we can move to an *autoregressive structure*. For example, in a first-order autoregressive specification, the correlation between time points r and s is assumed to be $\rho^{|r-s|}$. The necessary instruction for fitting the previously considered model but with this first-order autoregressive structure for the correlations is

    xtgee dep group pre visit, corr(ar1)

The output shown in Display 9.1 includes the note that eight subjects had to be excluded because they had fewer than two observations so that they could not contribute to the estimation of the correlation matrix. The estimates of the regression coefficients and their standard errors have changed, but not substantially. The estimated within-subject correlation matrix can be obtained using

```
note:  some groups have fewer than 2 observations
       not possible to estimate correlations for those groups
       8 groups omitted from estimation
Iteration 1: tolerance = 0.09563585
Iteration 2: tolerance = 0.00095295
Iteration 3: tolerance =  0.0000154
Iteration 4: tolerance =  2.500e-07
General estimating equation for panel data     Number of obs     =        287
Group and time vars:            subj visit     Number of groups  =         53
Link:                             identity     Obs/group, min    =          2
Family:                           Gaussian                  avg  =       5.42
Correlation:                         AR(1)                  max   =          6
                                                chi2(3)           =      64.31
Scale parameter:                 26.17334       Prob > chi2      =     0.0000
Pearson chi2(283):                7407.06       Deviance         =    7407.06
Dispersion (Pearson):            26.17334       Dispersion       =   26.17334
-----------------------------------------------------------------------------
    dep |      Coef.   Std. Err.       z     P>|z|      [95% Conf. Interval]
--------+--------------------------------------------------------------------
  group | -4.225908   1.040508     -4.061   0.000    -6.265266   -2.186551
    pre |  .4296511   .1359194      3.161   0.002     .1632538    .6960483
  visit | -1.188076   .1943142     -6.114   0.000    -1.568925   -.8072271
  _cons |  8.987358   3.002464      2.993   0.003     3.102637    14.87208
-----------------------------------------------------------------------------
```

Display 9.1

xtcorr

```
Estimated within-subj correlation matrix R:
        c1       c2       c3       c4       c5       c6
r1  1.0000
r2  0.6582   1.0000
r3  0.4332   0.6582   1.0000
r4  0.2852   0.4332   0.6582   1.0000
r5  0.1877   0.2852   0.4332   0.6582   1.0000
r6  0.1235   0.1877   0.2852   0.4332   0.6582   1.0000
```

which has the expected pattern in which correlations decrease substantially as the separation between the observations increases.

Other correlation structures are available using xtgee, including the correlation(unstructured) option in which no constraints are placed on the correlations. (This is essentially equivalent to multivariate analysis of variance for longitudinal data.) It might appear that using this option would be the most sensible one to choose for *all* data sets. This is not, however, the case since it necessitates the estimation of many nuisance parameters. This can, in some circumstances, cause problems in the estimation of those parameters of most interest, particularly when the sample size is small and the number of time points is large.

9.3.2 Chemotherapy data

We now analyize the chemotherapy data using a similar model as for the depression data, again using baseline, group and time variables as covariates, but using the Poisson distribution and log link. Assuming the data are available in an ASCII file called *chemo.dat*, they can be read into Stata in the usual way using

```
infile subj y1 y2 y3 y4 treat baseline age using chemo.dat
```

Some useful summary statistics can be obtained using

```
summarize y1 y2 y3 y4 if treat==0
```

Variable	Obs	Mean	Std. Dev.	Min	Max
y1	28	9.357143	10.13689	0	40
y2	28	8.285714	8.164318	0	29
y3	28	8.785714	14.67262	0	76
y4	28	7.964286	7.627835	0	29

```
summarize y1 y2 y3 y4 if treat==1
```

Variable	Obs	Mean	Std. Dev.	Min	Max
y1	31	8.580645	18.24057	0	102
y2	31	8.419355	11.85966	0	65
y3	31	8.129032	13.89422	0	72
y4	31	6.709677	11.26408	0	63

The largest value of *y1* in the progabide group seems out of step with the other maximum values and may indicate an outlier. Some graphics of the data can be useful for investigating this possibility further, but first it is convenient to transform the data from its present 'wide' form to the 'long' form. The necessary command is as follows:

```
reshape long y, i(subj) j(week)
sort subj treat week
list in 1/12
```

	subj	week	treat	baseline	age	y
1.	1	1	0	11	31	5
2.	1	2	0	11	31	3
3.	1	3	0	11	31	3
4.	1	4	0	11	31	3
5.	2	1	0	11	30	3
6.	2	2	0	11	30	5
7.	2	3	0	11	30	3
8.	2	4	0	11	30	3
9.	3	1	0	6	25	2
10.	3	2	0	6	25	4
11.	3	3	0	6	25	0
12.	3	4	0	6	25	5

Perhaps, the most useful graphical display for investigating the data is a set of graphs of individual response profiles. In order to include the baseline value in the graphs, duplicate the first observation for each subject (*week*=1) and change one of the duplicates for each subject to represent the baseline measure (week=0). Since the baseline measure represents seizure counts over an 8-week period, compared with 2-week periods for each of the other time-points, we divide the baseline measure by 4:

```
expand 2 if week==1
sort subj week
qui by subj: replace week=0 if _n==1
replace y=baseline/4 if week==0
```

Since we are planning to fit a Poisson model with the log link to the data, we take the log transformation before plotting the response profiles. In order to avoid too much overlap between subjects, we use nine graphs in each group. First, we produce a grouping variable *dum9* with nine levels:

```
gen ly = log(y+1)
sort treat baseline subj week
gen i=0
quietly by treat baseline subj: replace i=1 if _n==1
replace i=sum(i)
gen dum9=mod(i,9)
```

Sorting by baseline before placing sets of consecutive numbers from 1 to 9 into *dum9* ensures that the set of baseline measures for each graph are not too close together. We can now produce the graphs:

```
sort dum9 subj week
graph ly week if treat==0, c(L) s([subj]) by(dum9) /*
       */ l1(" ") bs(5) b1("placebo group") b2(" ")
graph ly week if treat==1, c(L) s([subj]) by(dum9) /*
       */ l1(" ") bs(5) b1("progabide group") b2(" ")
drop if week==0
```

The resulting graphs are shown in Figures 9.1 and 9.2. There is no obvious improvement in the progabide group. Subject 49 had more epileptic fits overall than any other subject. However, since we are going to control for the baseline measure, it is an unusual shape (not the overall level) of the response profile that makes a subject an outlier.

As discussed in Chapter 7, the most plausible distribution for count data is the Poisson distribution. The Poisson distribution is introduced into xtgee models by specifying family(poisson). The log link is implied (since it is the canonical link) and the logarithm of number of seizures is regressed on the chosen covariates. The summary tables for the seizure data given above provide strong empirical evidence that there is overdispersion and this can be incorporated using the scale(x2) option to allow the dispersion parameter ϕ to be estimated (see Chapter 7).

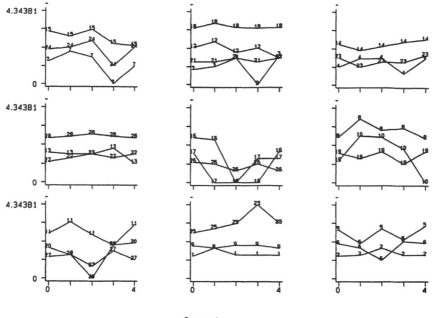

placebo group

Figure 9.1 *Response profiles in placebo group.*

```
xtgee y treat baseline age week, corr(exc) family(pois) scale(x2)
```

The output, assuming an exchangeable correlation structure, is given in Display 9.2 and the estimated correlation matrix is obtained using xtcorr.

```
Estimated within-subj correlation matrix R:
        c1       c2       c3       c4
r1  1.0000
r2  0.4033  1.0000
r3  0.4033  0.4033  1.0000
r4  0.4033  0.4033  0.4033  1.0000
```

In Display 9.2, the parameter ϕ is estimated to be 5.11, indicating the severe overdispersion in these data. We briefly illustrate how important it is to allow for overdispersion by omitting the scale(x2) option:

```
xtgee y treat baseline age week, corr(exc) family(pois)
```

The results given in Display 9.3 show that the variable *treat* becomes significant, leading to a qualitatively different conclusion about the effectiveness of the treatment. Even if overdispersion had not been suspected, this error could have been detected by using the robust option (see Chapter 7):

```
Iteration 1: tolerance = 0.0183008
Iteration 2: tolerance = 2.535e-06
Iteration 3: tolerance = 1.030e-09

General estimating equation for panel data    Number of obs    =       236
Group variable:                        subj    Number of groups =        59
Link:                                   log    Obs/group, min   =         4
Family:                             Poisson                 avg   =      4.00
Correlation:                    exchangeable                max   =         4
                                               chi2(4)          =    189.98
Scale parameter:                   5.107904    Prob > chi2      =    0.0000
Pearson chi2(231):                 1179.93     Deviance         =    950.13
Dispersion (Pearson):              5.107904    Dispersion       =  4.113122

--------------------------------------------------------------------------
       y |     Coef.   Std. Err.       z    P>|z|     [95% Conf. Interval]
---------+----------------------------------------------------------------
   treat | -0.1478458  0.1604068   -0.922   0.357   -0.4622373   0.1665457
baseline |  0.0227431  0.001708    13.316   0.000    0.0193956   0.0260906
     age |  0.0235715  0.0135012    1.746   0.081   -0.0028903   0.0500334
    week | -0.0587233  0.0354633   -1.656   0.098   -0.12823     0.0107834
   _cons |  0.6759401  0.4630713    1.460   0.144    0.231663    1.683643
--------------------------------------------------------------------------
(Standard errors scaled using square root of Pearson X2-based dispersion)
```

Display 9.2

```
Iteration 1: tolerance = 0.0183008
Iteration 2: tolerance = 2.535e-06
Iteration 3: tolerance = 1.030e-09

General estimating equation for panel data    Number of obs    =       236
Group variable:                        subj    Number of groups =        59
Link:                                   log    Obs/group, min   =         4
Family:                             Poisson                 avg   =      4.00
Correlation:                    exchangeable                max   =         4
                                               chi2(4)          =    970.41
Scale parameter:                          1    Prob > chi2      =    0.0000
Pearson chi2(231):                 1179.93     Deviance         =    950.13
Dispersion (Pearson):              5.107904    Dispersion       =  4.113122

--------------------------------------------------------------------------
       y |     Coef.   Std. Err.       z    P>|z|     [95% Conf. Interval]
---------+----------------------------------------------------------------
   treat | -.1478458   .0709743    -2.083   0.037   -.286953    -.0087386
baseline |  .0227431   .0007557    30.095   0.000    .021262     .0242243
     age |  .0235715   .0059738     3.946   0.000    .0118631    .03528
    week | -.0587233   .0156912    -3.742   0.000   -.0894770   -.0279691
   _cons |  .6759401   .2048927     3.299   0.001    .2743578   1.077522
--------------------------------------------------------------------------
```

Display 9.3

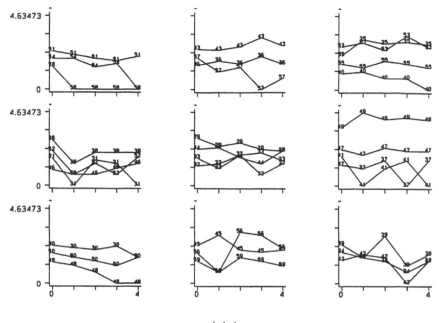

progabide group

Figure 9.2 *Response profiles in the treated group.*

```
xtgee y treat baseline age week, corr(exc) family(pois) robust
```

The results of the robust regression in Display 9.4 are remarkably similar to the model-based standard error of the model allowing for overdispersion, indicating that the latter is a reasonable model for the data.

The estimated coefficient of *treat* describes the difference in the log of the average seizure count between the placebo and prograbide treated groups. A negative value indicates that the treatment is effective relative to the placebo in controlling the seizure rate although this is not significant. The exponentiated coefficient gives an incidence rate ratio; here it represents the ratio of average seizure rates, measured as the number of seizures per 2-week period, for the treated patients compared to that among the control patients. The exponentiated coefficient and the corresponding confidence interval can be obtained directly using the eform option in xtgee:

```
xtgee y treat baseline age week, corr(exc) family(pois) scale(x2)
    eform
```

The results in Display 9.5 indicate that there is a 14% reduction in the incidence of epileptic seizures in the treated group compared with the control

```
Iteration 1: tolerance = 0.0183008
Iteration 2: tolerance = 2.535e-06
Iteration 3: tolerance = 1.030e-09
```

General estimating equation for panel data		Number of obs	=	236
Group variable:	subj	Number of groups	=	59
Link:	log	Obs/group, min	=	4
Family:	Poisson	avg	=	4.00
Correlation:	exchangeable	max	=	4
		chi2(4)	=	603.40
Scale parameter:	1	Prob > chi2	=	0.0000
Pearson chi2(231):	1179.93	Deviance	=	950.13
Dispersion (Pearson):	5.107904	Dispersion	=	4.113122

(standard errors adjusted for clustering on subj)

y	Coef.	Robust Std. Err.	z	P>|z|	[95% Conf. Interval]	
treat	-.1478458	.1701226	-0.869	0.385	-.4812799	.1855883
baseline	.0227431	.0012543	18.133	0.000	.0202848	.0252014
age	.0235715	.0119041	1.980	0.048	.0002399	.0460031
week	-.0587233	.035298	-1.664	0.096	-.1279062	.0104595
_cons	.6759401	.3570996	1.893	0.058	-.0239622	1.375842

Display 9.4

```
Iteration 1: tolerance = 0.0183008
Iteration 2: tolerance = 2.535e-06
Iteration 3: tolerance = 1.031e-09
```

General estimating equation for panel data		Number of obs	=	236
Group variable:	subj	Number of groups	=	59
Link:	log	Obs/group, min	=	4
Family:	Poisson	avg	=	4.00
Correlation:	exchangeable	max	=	4
		chi2(4)	=	189.98
Scale parameter:	5.107904	Prob > chi2	=	0.0000
Pearson chi2(231):	1179.93	Deviance	=	950.13
Dispersion (Pearson):	5.107904	Dispersion	=	4.113122

y	IRR	Std. Err.	z	P>|z|	[95% Conf. Interval]	
treat	.8625641	.1383611	-0.922	0.357	.6298728	1.181218
baseline	1.023004	.0017472	13.316	0.000	1.019585	1.026434
age	1.023852	.0138232	1.740	0.081	.9971138	1.051306
week	.9429676	.0334407	-1.656	0.098	.879651	1.010842

(Standard errors scaled using square root of Pearson X2-based dispersion)

Display 9.5

group. According to the confidence interval, the reduction could be as low as 38% or there could be as much as an 18% increase.

When these data were analyzed by Thall and Vail (1990), a possible interaction between baseline seizure count and treatment group was allowed for. Such a model is easily fitted using the following instructions:

```
gen blint = treat*baseline
xtgee y treat baseline age week blint, /*
        */corr(exc) scale(x2) family(pois) eform
```

```
Iteration 1: tolerance = 0.01738938
Iteration 2: tolerance = 0.00001068
Iteration 3: tolerance = 1.147e-08

General estimating equation for panel data    Number of obs   =        236
Group variable:                       subj    Number of groups =        59
Link:                                  log    Obs/group, min  =          4
Family:                            Poisson                avg  =       4.00
Correlation:                 exchangeable                 max  =          4
                                               chi2(5)         =     190.86
Scale parameter:                  5.151199     Prob > chi2     =     0.0000
Pearson chi2(230):                1184.78      Deviance        =     947.21
Dispersion (Pearson):             5.151199     Dispersion      =   4.118309


-----------------------------------------------------------------------------
      y |       IRR   Std. Err.       z    P>|z|     [95% Conf. Interval]
--------+--------------------------------------------------------------------
  treat | 0.7735594  0.1996358    -0.995   0.320    0.4664664    1.282824
baseline| 1.021598   0.0031575     6.914   0.000    1.015428     1.027806
    age | 1.025602   0.0142751     1.816   0.069    0.9980015    1.053966
   week | 0.9429676  0.0335678    -1.650   0.099    0.8794186    1.011109
  blint | 1.002      0.0037218     0.538   0.591    0.9947316    1.009321
-----------------------------------------------------------------------------

(Standard errors scaled using square root of Pearson X2-based dispersion)
```

Display 9.6

There is no evidence of an interaction in Display 9.6. Other correlation structures might be explored for these data (see exercises).

We now look at the standardized Pearson residuals of the previous model separately for each week, as an informal method for finding outliers (see equation (7.12)). This can be done using the predict command to obtain estimated counts and computing the standardized Pearson residuals from them.

```
xtgee y treat baseline age week, corr(exc) family(pois) scale(x2)
predict pred, xb
replace pred = exp(pred)
gen pres = (y-pred)/sqrt(pred)
gen stpres = pres/sqrt($S_E_P2)
sort week
graph stpres, box s([subj]) by(week) ylab
```

The resulting graph is shown in Figure 9.3

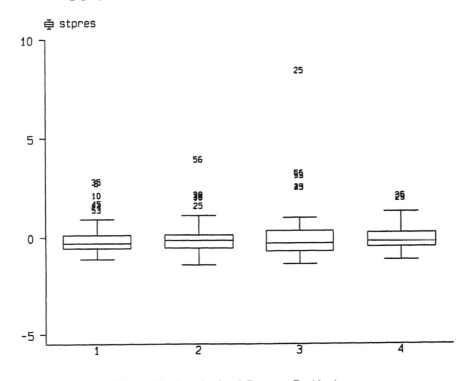

Figure 9.3 *Standardized Pearson Residuals.*

Here, subject 25 is shown to be an outlier in week 3. In Figure 9.1 it is apparent that this is due to a large increase in the log of the seizure counts at week 3.

9.4 Exercises

1. Run the random effects model using different estimation methods available for **xtreg**.

2. Compare the result of the above with the result of running ordinary linear regression with standard errors corrected for the within-subject correlation using

   a. The option, **robust, cluster(subj)**, see help for **regress**;

   b. Bootstrapping, by sampling subjects with replacement, not observations. This can be achieved using the **bs** command, together with the option **cluster(subj)**.

3. Explore other possible correlational structures for the seizure data in the

context of a Poisson model. Examine the robust standard errors in each case.

4. Investigate what Poisson models are most suitable when subject 25 is excluded from the analysis.

5. Repeat the analysis treating the baseline measure (divided by four) as another response variable and replacing the variable *week* by a simple indicator variable, *post*, which is 0 at baseline and 1 subsequently. Is there a significant treatment effect, now represented by the interaction between *post* and *treat*?

CHAPTER 10

Some Epidemiology

10.1 Description of data

This chapter illustrates a number of different problems in epidemiology using four datasets that are presented in the form of cross-tabulations in Table 10.1. The first table (a) is taken from Clayton and Hills (1993). The data result from cohort study that was carried out to investigate the relationship between energy intake and ischemic heart disease (IHD). Low energy intake indicates lack of physical exercise and is therefore considered a risk factor. The table gives frequencies of IHD by 10-year age-band and exposure to a high or low calorie diet. The total person-years of observation are also given for each cell.

The second dataset (b) is the result of a case-control study investigating whether keeping a pet bird is a risk factor for lung cancer. This dataset is given in Hand et al. (1994).

The last two datasets (c) and (d) are from matched case control studies. In dataset (c), cases of endometrial cancer were matched on age, race, date of admission, and hospital of admission to a suitable control not suffering from cancer, and past exposure to conjugated estrogens was determined. The dataset is described in Everitt (1994). The last set of data, described in Clayton and Hills (1993), arises from a case-control study of breast cancer screening. Women who had died of breast cancer were matched with three control women. The screening history of the subjects in each matched case-control set was assessed over the period up to the time of diagnosis of the case.

10.2 Introduction to epidemiology

Epidemiology is the study of diseases in populations, in particular the search for causes of disease. For ethical reasons, subjects cannot be randomized to possible risk factors in order to establish whether these are associated with an increase in the incidence of disease. Instead, epidemiology is based on observational studies, the most important types being cohort studies and case-control studies. We will give a very brief description of the design and analysis of these two types of studies, following closely the explanations and notation given in the excellent book, *Statistical Models in Epidemiology* by Clayton and Hills (1993).

In a cohort study, a group of subjects free of the disease is followed up and the presence of risk factors as well as the occurrence of the disease of interest is recorded. This design is illustrated in Figure 10.1. An example of a cohort study

Table 10.1 Simple tables in epidemiology (a) number of IHD cases and person-years of observation by age and exposure to low energy diet. (b) The number of lung cancer cases and controls who keep a pet bird (c) frequency of exposure to oral conjugated estrogens among cases of endometrial cancer and their matched controls (d) Screening history in subjects who died of breast cancer and 3 matched controls.
(Tables (a) & (d) taken from Clayton and Hills (1993) with permission of their publisher, Oxford University Press)

(a) Cohort Study

Age	Exposed < 2750 kcal		Unexposed ≥ 2750 kcal	
	Cases	Pers-yrs	Cases	Pers-yrs
40-49	2	311.9	4	607.9
50-59	12	878.1	5	1272.1
60-69	14	667.5	8	888.9

(b) Case-Control Study

Kept pet birds	Cases	Controls
yes	98	101
no	141	328
Total	239	429

(c) 1:1 Matched Case-Control Study

Cases		Controls		
		+	−	Total
	+	12	43	55
	−	7	121	128
	Total	19	164	183

(d) 1:3 Matched Case-Control Study

Status of the case	Number of controls screened			
	0	1	2	3
Screened	1	4	3	1
Unscreened	11	10	12	4

is the study described in the previous section where subjects were followed up in order to monitor the occurrence of ischemic heart disease in two risk groups, those with high and low energy intake, giving the results in Table 10.1(a). The incidence rate λ (of the disease) can be estimated by the number of new cases of the disease D during a time interval divided by the person-time of observation Y, the sum of all subjects' periods of observation during the time interval:

$$\hat{\lambda} = \frac{D}{Y} .$$
(10.1)

This is the maximum likelihood estimator of λ assuming that D follows the Poisson distribution (independent events occurring at a constant probability rate in continuous time) with mean λY. The most important quantity of interest in a cohort study is the incidence rate ratio (or relative risk) between subgroups of subjects defined by their exposure to the risk factor. This may be estimated by

$$\hat{\theta} = \frac{D_1/Y_1}{D_0/Y_0}$$
(10.2)

where the subscripts 1 and 0 denote exposed and unexposed groups, respec-

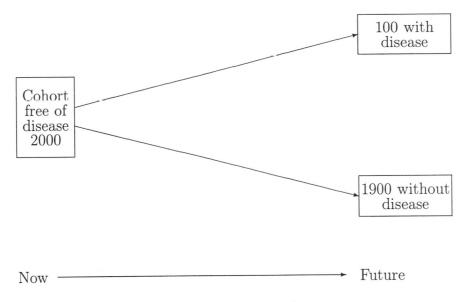

Figure 10.1 *Cohort study.*

tively. This estimator can be derived by maximizing the conditional likelihood that these were D_1 cases in the exposed group conditional on these being a total of $D = D_0 + D_1$ cases. Since this conditional likelihood for θ is a true likelihood based on the binomial distribution, exact confidence intervals can be obtained. However, a potential problem in estimating this rate ratio is confounding arising from systematic differences in prognostic factors between the groups. This problem can be dealt with by dividing the cohort into groups or strata according to such prognostic factors and assuming that the rate ratio for exposed and unexposed subjects is the same in each stratum. If there are D^t cases and Y^t years of observation in stratum t, then the common rate ratio can be estimated using the method of Mantel and Haenszel as follows:

$$\hat{\theta} = \frac{\sum_t D_1^t Y_0^t / Y^t}{\sum_t D_0^t Y_1^t / Y^t} . \tag{10.3}$$

Note that the strata might not correspond to groups of subjects. For example, if the confounder is age, subjects who cross from one age-band into the next during the study, contribute parts of their periods of observation to different strata. This is how Table 10.1(a) was constructed. Another way of controlling for confounding variables is to use Poisson regression to model the number of occurrences of disease or failures. If a log link is used, the expected number of failures can be made proportional to the person-years of observation by adding the log of the person-years of observation to the linear predictor as an offset, giving

$$\log(D) = \log(Y) + \boldsymbol{\beta}^T \mathbf{x} \tag{10.4}$$

Exponentiating the equation gives

$$D = Y \exp\left(\boldsymbol{\beta}^T \mathbf{x}\right) \tag{10.5}$$

as required.

If the incidence rate of a disease is small, a cohort study requires a large number of person-years of observation, making it very expensive. A more feasible type of study in this situation is a case-control study in which cases of the disease of interest are compared with non-cases, called controls, with respect to exposure to possible risk factors in the past. A diagram illustrating the basic idea of case-control studies is shown in Figure 10.2. The assumption here is that

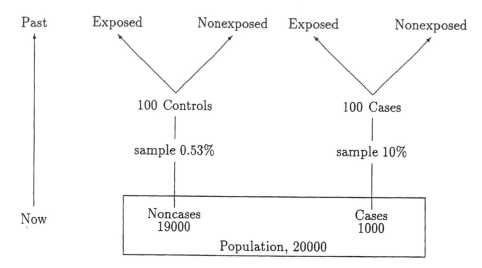

Figure 10.2 *Case-control study.*

the probability of selection into the study is independent of the exposures of interest. The data in Table 10.1(b) derive from a case-control study in which cases with lung cancer and healthy controls were interviewed to ascertain whether or not they had been "exposed" to a pet bird.

Let D and H be the number of cases and controls, respectively, and let the subscripts 0 and 1 denote "unexposed" and "exposed." Since the proportion of cases was determined by the design, it is not possible to estimate the relative risk of disease between exposed and nonexposed subjects. However, the odds of exposure in the cases or controls can be estimated and the ratio of these odds

$$\frac{D_1/D_0}{H_1/H_0} = \frac{D_1/H_1}{D_0/H_0} \tag{10.6}$$

is equal to to the odds ratio of being a case in the exposed group compared with the unexposed group. If we model the (log) odds of being a case using logistic regression with the exposure as an explanatory variable, then the coefficient of the exposure variable is an estimate of the true log odds ratio although the estimate of the odds itself (reflected by the constant) only reflects the proportion of cases in the study. Logistic regression can therefore be used to control for confounding variables.

A major difficulty with case-control studies is to find suitable controls who are similar enough to the cases (so that differences in exposure can reasonably be assumed to be due to their association with the disease) without being overmatched, which can result in very similar exposure patterns. The problem of finding cases who are similar enough is often addressed by matching cases individually to controls according to important variables such as age and sex. Examples of such matched case-control studies are given in Table 10.1(b) and (c). In the screening study, matching had the following additional advantage noted in Clayton and Hills (1983). The screening history of controls could be determined by considering only the period up to the diagnosis of the case, ensuring that cases did not have a decreased opportunity for screening because they would not have been screened after their diagnosis. The statistical analysis must take into account the matching. Two methods of analysis are McNemar's test in the case of two by two tables and conditional logistic regression in the case of several controls per case and/or several explanatory variables.

Since the case-control sets have been matched on variables that are believed to be associated with disease status, the sets can be thought of as strata with subjects in one stratum having higher or lower odds of being a case than those in another after controlling for the exposures. A logistic regression model would have to accommodate these differences by including a parameter ω_c for each case-control set so that the log odds of being a case for subject i in case-control set c would be

$$\log\left(\omega_{ci}\right) = \log\left(\omega_c\right) + \boldsymbol{\beta}^T \mathbf{x}_i \qquad (10.7)$$

However, this would result in too many parameters to be estimated. Furthermore, the parameters ω_c are of no interest to us. (One obvious solution to the problem would be to treat $\log(\omega_c)$ as a random effect, giving a random intercept model, see Chapter 9.) In conditional logistic regression, these nuisance parameters are eliminated by making use of the fact that there is only one case in each case-control set. For example, in a 1:1 matched case-control study, ignoring the fact that each set has one case, the probability that subject 1 in the set is a case and subject 2 is a noncase is

$$\Pr(1) = \frac{\omega_{c1}}{1 + \omega_{c1}} \times \frac{1}{1 + \omega_{c2}} \qquad (10.8)$$

and the probability that subject 1 is a noncase and subject 2 is a case is

$$\Pr(2) = \frac{1}{1 + \omega_{c1}} \times \frac{\omega_{c2}}{1 + \omega_{c2}} \tag{10.9}$$

However, conditional on there being one case in a set, the probability of subject 1 being the case is simply

$$\frac{\Pr(1)}{\Pr(1) + \Pr(2)} = \omega_{c1} / (\omega_{c1} + \omega_{c2}) = \frac{\exp\left(\beta^T \mathbf{x}_1\right)}{\exp\left(\beta^T \mathbf{x}_1\right) + \exp\left(\beta^T \mathbf{x}_2\right)} \tag{10.10}$$

since ω_c cancels out, see equation (10.7). The expression on the right-hand side of (10.10) is the contribution of a single case-control set to the conditional likelihood of the sample. Similarly, it can be shown that if there are k controls per case and the subjects within each case-control set are labeled 1 for the case and 2 to $k + 1$ for the controls then the log-likelihood becomes

$$\sum_c \log \left(\frac{\exp\left(\beta^T \mathbf{x}_1\right)}{\sum_{i=2}^{k+1} \exp\left(\beta^T \mathbf{x}_i\right)} \right) \tag{10.11}$$

10.3 Analysis using Stata

There is a collection of instructions in Stata, the `epitab` commands, that can be used to analyze small tables in epidemiology. These commands either refer to variables in an existing dataset or they can take cell counts as arguments (i.e., there are immediate commands, see Chapter 1).

The first dataset (a) is given in a file as tabulated and can be read using

```
infile str5 age num0 py0 num1 py1 using ihd.dat,clear
gen agegr=_n
reshape long num py, i(agegr) j(exposed)
```

Ignoring age, the incidence rate ratio can be estimated using

```
ir num exposed py
```

```
                            exposed
                     |  Exposed   Unexposed  |    Total
---------------------+-------------------------+----------
              num    |     17          28    |      45
              py     |    2769        1857    |     4626
---------------------+-------------------------+----------
                     |                         |
   Incidence Rate    |  .0061394    .0150781  |  .0097276
                     |                         |
                     |         Pt. Est.        |  [95% Conf. Interval]
                     |-------------------------+----------------------
   Inc. rate diff.   |        -.0089387        |  -.0152401   -.0026372
   Inc. rate ratio   |         .4071738        |   .2090942    .7703146  (exact)
   Prev. frac. ex.   |         .5928262        |   .2296854    .7909058  (exact)
   Prev. frac. pop   |         .3548499        |
                     +--------------------------------------------------
                       (midp)   Pr(k<=17) =                 0.0016  (exact)
                       (midp)  2*Pr(k<=17) =                0.0031  (exact)
```

with the result that the incidence rate ratio is estimated as 0.41 with a 95% confidence interval from 0.21 to 0.77. The terms (exact) imply that that the confidence intervals are exact (no approximation was used). Controlling for age using the epitab command

```
ir num exposed py, by(age)
```

```
          age  |     IRR      [95% Conf. Interval]    M-H Weight
---------------+------------------------------------------------
        40-49  |  1.026316    .1470886   11.3452      1.321739   (exact)
        50-59  |   .2876048   .0793725    .8770737    7.099535   (exact)
        60-69  |   .4293749   .1560738   1.096694     7.993577   (exact)
---------------+------------------------------------------------
        Crude  |   .4071738   .2090942    .7703146                (exact)
 M-H combined  |   .4161246   .2262261    .7654276
---------------+------------------------------------------------
 Test for heterogeneity (M-H)    chi2(2) =      1.57  Pr>chi2 = 0.4554
```

gives very similar estimates, as shown in the row labeled "M-H combined" (the Mantel Haentzel estimate). Another way of controlling for age is to carry out Poisson regression with the log of *py* as an offset, using

```
poisson num exposed, e(py) irr
```

```
Iteration 0: Log Likelihood = -14.24379
Iteration 1: Log Likelihood = -13.907623
Iteration 2: Log Likelihood = -13.906418

Poisson regression, normalized by py          Number of obs   =        6
Goodness-of-fit chi2(4)       =      5.689    Model chi2(1)   =    8.889
Prob > chi2                   =     0.2236    Prob > chi2     =   0.0029
Log Likelihood                =    -13.906    Pseudo R2       =   0.2422

-----------------------------------------------------------------------
    num |      IRR   Std. Err.      z    P>|z|     [95% Conf. Interval]
--------+--------------------------------------------------------------
exposed |  .4072981    .125232   -2.921   0.003     .2229428    .7441002
-----------------------------------------------------------------------
```

showing that there is an incidence rate ratio of 0.41 with a 95% confidence interval from 0.22 to 0.74. Another way of achieving the same result is using glm as follows:

```
gen lpy=log(py)
glm num exposed, fam(poisson) link(log) off(lpy) eform
```

We will analyize the case-control study using the "immediate" command cci. According to help epitab, the following notation is used for immediate commands.

```
                        Exposed    Unexposed
            ------------+----------------------
            Cases       |    a          b
            Noncases    |    c          d
                          Exposed    Unexposed
            ------------+----------------------
            Cases       |    a          b
            Person-time |    N1         N0
```

The quantities a, b, etc. in the table are used in lexicographic order, i.e., a, b, c, d, for case-control studies or cohort studies with equal follow-up and a, b, N1, N0 for cohort studies with nonequal follow-up.

The bird data can therefore be analyzed using the immediate command for case-control studies, cci, as follows:

```
cci 98 141 101 328
```

```
                                                         Proportion
                  |  Exposed   Unexposed |    Total     Exposed
------------------+----------------------+----------------------------
         Cases |      98        141    |     239      0.4100
         Controls |    101        328    |     429      0.2354
------------------+----------------------+----------------------------
         Total |     199        469    |     668      0.2979
                  |                      |
                  |     Pt. Est.         |  [95% Conf. Interval]
                  |----------------------+----------------------------
      Odds ratio |     2.257145         |  1.606121    3.172163  (Cornfield)
   Attr. frac. ex. |     .5569624         |   .3773821    .6847577  (Cornfield)
   Attr. frac. pop |     .2283779         |
                  +----------------------------------------------------
                        chi2(1) =     22.37  Pr>chi2 = 0.0000
```

There is therefore a significant association between keeping pet birds and developing lung cancer. The word (Cornfield) is there to remind us that an approximation, the Cornfield approximation, was used to estimate the confidence interval (see, for example, Rothman, 1986).

The matched case-control study with one control per case may be analyzed using the immediate command mcci where the columns of the two-way cross-tabulation represent exposed and unexposed controls:

mcci 12 43 7 121

```
                             Controls
      Cases           |  Exposed   Unexposed |    Total
----------------------+----------------------+----------
            Exposed |     12         43    |     55
          Unexposed |      7        121    |    128
----------------------+----------------------+----------
             Total |     19        164    |    183

      McNemar's chi2(1) =     25.92        Pr>chi2 = 0.0000

      Proportion with factor
            Cases        .3005464
            Controls     .1038251          [95% conf. interval]
                         ---------          --------------------
            difference   .1967213           .1210924    .2723502
            ratio        2.894737           1.885462    4.444269
            rel. diff.   .2195122           .1448549    .2941695

            odds ratio   6.142857           2.739803    16.18481  (exact)
```

The matched case-control study with three controls per case cannot be analyzed using epitab. Instead, we will use conditional logistic regression. We need to convert the data from the table shown in 10.1(d) into the form required for conditional logistic regression; that is, one observation per subject (including cases and controls); an indicator variable, *cancer*, for cases; another indicator variable, *screen*, for screening; and a third variable, *caseid*, giving the id for each case-control set of four women.

First, read the data and transpose them so that the columns in the transposed dataset have variable names *ncases0* and *ncases1*. The rows in this dataset correspond to different numbers of screened-matched controls, which are called *nconstr*. Then reshape to long, generating an indicator, *casescr*, for whether or not the case was screened:

```
infile v1-v4 using screen.dat,clear
gen str8 _varname="ncases1" in 1
replace _varname="ncases0" in 2
xpose, clear
gen nconscr=_n-1
reshape long ncases, i(nconscr) j(casescr)
list
```

	nconscr	casescr	ncases
1.	0	0	11
2.	0	1	1
3.	1	0	10
4.	1	1	4
5.	2	0	12
6.	2	1	3
7.	3	0	4
8.	3	1	1

The next step is to replicate each of the records *ncases* times so that we have one record per case-control set. Then define the variable *caseid* and expand the dataset four times in order to have one record per subject. The four subjects within each case-control are arbitrarily labeled 0 to 3 in the variable `control`, where 0 stands for "the case" and 1, 2, and 3 for the controls.

```
expand ncases
sort casescr nconscr
gen caseid=_n
expand 4
sort caseid
quietly by caseid: gen control=_n-1
list in 1/8
```

	nconscr	casescr	ncases	caseid	control
1.	0	0	11	1	0
2.	0	0	11	1	1
3.	0	0	11	1	2
4.	0	0	11	1	3
5.	0	0	11	2	0
6.	0	0	11	2	1
7.	0	0	11	2	2
8.	0	0	11	2	3

Now, *screen*, the indicator whether a subject was screened, is defined to be zero except for the cases who were screened and for as many controls as were screened according to *nconscr*. The variable *cancer* is one for cases and zero otherwise.

```
gen screen=0
replace screen=1 if control==0&casescr==1   /* the case */
replace screen=1 if control==1&nconscr>0
replace screen=1 if control==2&nconscr>1
replace screen=1 if control==3&nconscr>2
gen cancer=0
replace cancer=1 if control==0
```

We can reproduce Table 10.1(d) by temporarily collapsing the data (using preserve and restore to revert back to the original data) as follows:

```
preserve
collapse (sum) screen (mean) casescr , by(caseid)
gen nconscr=screen-casescr
tab casescr nconscr
restore
```

(mean) casescr	nconscr 0	1	2	3	Total
0	11	10	12	4	37
1	1	4	3	1	9
Total	12	14	15	5	46

Now, we are ready to carry out conditional logistic regression:

```
clogit cancer screen, group(caseid) or
```

```
Iteration 0:  Log Likelihood =-62.404527
Iteration 1:  Log Likelihood =-59.212727
Iteration 2:  Log Likelihood = -59.18163
Iteration 3:  Log Likelihood =-59.181616

Conditional (fixed-effects) logistic regression      Number of obs =     184
                                                      chi2(1)       =    9.18
                                                      Prob > chi2   = 0.0025
Log Likelihood = -59.181616                           Pseudo R2     = 0.0719

------------------------------------------------------------------------------
  cancer | Odds Ratio   Std. Err.      z    P>|z|    [95% Conf. Interval]
---------+--------------------------------------------------------------------
  screen |  .2995581    .1278367   -2.825   0.005    .1297866     .6914043
------------------------------------------------------------------------------
```

Screening is therefore protective of death from breast cancer, reducing the odds to a third (95% 0.13 to 0.69).

10.4 Exercises

1. Carry out conditional logistic regression to estimate the odds ratio for the data in Table 10.1(c). The data are given in the same form as in the Table in a file called *estrogen.dat*.

2. For the data *ihd. dat*, use the command `iri` to calculate the incidence rate ratio for IHD without controlling for age.

3. Use Poisson regression to test whether the effect of exposure on incidence of IHD differs significantly between age groups.

4. Repeat the analysis of exercise 5 in Chapter 9, using the raw seizure counts (without dividing the baseline count by 4) as dependent variables and the log of the observation time (8 weeks for baseline and 2 weeks for all other counts) as an offset.

CHAPTER 11

Survival Analysis: Retention of Heroin Addicts in Methadone Maintenance Treatment

11.1 Description of data

The "survival data" to be analyzed in this chapter are the times that heroin addicts remained in a clinic for methadone maintenance treatment. Here, the endpoint of interest is not death as the word "survival" suggests, but termination of treatment. Some subjects were still in the clinic at the time these data were recorded and this is indicated by the variable *status*, which is equal to 1 if the person departed and 0 otherwise. Possible explanatory variables for retention in treatment are maximum methadone dose and a prison record, as well as which of two clinics the addict was treated in. These variables are called *dose*, *prison*, and *clinic*, respectively. The data were first analyzed by Caplehorn and Bell (1991) and also appear in Hand et al. (1994).

Table 11.1 Data in *heroin.dat*

id	clinic	status	time	prison	dose	id	clinic	status	time	prison	dose
1	1	1	428	0	50	132	2	0	633	0	70
2	1	1	275	1	55	133	2	1	661	0	40
3	1	1	262	0	55	134	2	1	232	1	70
4	1	1	183	0	30	135	2	1	13	1	60
5	1	1	259	1	65	137	2	0	563	0	70
6	1	1	714	0	55	138	2	0	969	0	80
7	1	1	438	1	65	143	2	0	1052	0	80
8	1	0	796	1	60	144	2	0	944	1	80
9	1	1	892	0	50	145	2	0	881	0	80
10	1	1	393	1	65	146	2	1	190	1	50
11	1	0	161	1	80	148	2	1	79	0	40
12	1	1	836	1	60	149	2	0	884	1	50
13	1	1	523	0	55	150	2	1	170	0	40
14	1	1	612	0	70	153	2	1	286	0	45
15	1	1	212	1	60	156	2	0	358	0	60
16	1	1	399	1	60	158	2	0	326	1	60
17	1	1	771	1	75	159	2	0	769	1	40
18	1	1	514	1	80	160	2	1	161	0	40
19	1	1	512	0	80	161	2	0	564	1	80
21	1	1	624	1	80	162	2	1	268	1	70
22	1	1	209	1	60	163	2	0	611	1	40
23	1	1	341	1	60	164	2	1	322	0	55
24	1	1	299	0	55	165	2	0	1076	1	80
25	1	0	826	0	80	166	2	0	2	1	40
26	1	1	262	1	65	168	2	0	788	0	70
27	1	0	566	1	45	169	2	0	575	0	80

Table 11.1 Data in *heroin.dat*

28	1	1	368	1	55	170	2	1	109	1	70
30	1	1	302	1	50	171	2	0	730	1	80
31	1	0	602	0	60	172	2	0	790	0	90
32	1	1	652	0	80	173	2	0	456	1	70
33	1	1	293	0	65	175	2	1	231	1	60
34	1	0	564	0	60	176	2	1	143	1	70
36	1	1	394	1	55	177	2	0	86	1	40
37	1	1	755	1	65	178	2	0	1021	0	80
38	1	1	591	0	55	179	2	0	684	1	80
39	1	0	787	0	80	180	2	1	878	1	60
40	1	1	739	0	60	181	2	1	216	0	100
41	1	1	550	1	60	182	2	0	808	0	60
42	1	1	837	0	60	183	2	1	268	1	40
43	1	1	612	0	65	184	2	0	222	0	40
44	1	0	581	0	70	186	2	0	683	0	100
45	1	1	523	0	60	187	2	0	496	0	40
46	1	1	504	1	60	188	2	1	389	0	55
48	1	1	785	1	80	189	1	1	126	1	75
49	1	1	774	1	65	190	1	1	17	1	40
50	1	1	560	0	65	192	1	1	350	0	60
51	1	1	160	0	35	193	2	0	531	1	65
52	1	1	482	0	30	194	1	0	317	1	50
53	1	1	518	0	65	195	1	0	461	1	75
54	1	1	683	0	50	196	1	1	37	0	60
55	1	1	147	0	65	197	1	1	167	1	55
57	1	1	563	1	70	198	1	1	358	0	45
58	1	1	646	1	60	199	1	1	49	0	60
59	1	1	899	0	60	200	1	1	457	1	40
60	1	1	857	0	60	201	1	1	127	0	20
61	1	1	180	1	70	202	1	1	7	1	40
62	1	1	452	0	60	203	1	1	29	1	60
63	1	1	760	0	60	204	1	1	62	0	40
64	1	1	496	0	65	205	1	0	150	1	60
65	1	1	258	1	40	206	1	1	223	1	40
66	1	1	181	1	60	207	1	0	129	1	40
67	1	1	386	0	60	208	1	0	204	1	65
68	1	0	439	0	80	209	1	1	129	1	50
69	1	0	563	0	75	210	1	1	581	0	65
70	1	1	337	0	65	211	1	1	176	0	55
71	1	0	613	1	60	212	1	1	30	0	60
72	1	1	192	1	80	213	1	1	41	0	60
73	1	0	405	0	80	214	1	0	543	0	40
74	1	1	667	0	50	215	1	0	210	1	50
75	1	0	905	0	80	216	1	1	193	1	70
76	1	1	247	0	70	217	1	1	434	0	55
77	1	1	821	0	80	218	1	1	367	0	45
78	1	1	821	1	75	219	1	1	348	1	60
79	1	0	517	0	45	220	1	0	28	0	50
80	1	0	346	1	60	221	1	0	337	0	40
81	1	1	294	0	65	222	1	0	175	1	60
82	1	1	244	1	60	223	2	1	149	1	80
83	1	1	95	1	60	224	1	1	546	1	50
84	1	1	376	1	55	225	1	1	84	0	45
85	1	1	212	0	40	226	1	0	283	1	80
86	1	1	96	0	70	227	1	1	533	0	55
87	1	1	532	0	80	228	1	1	207	1	50
88	1	1	522	1	70	229	1	1	216	1	50
89	1	1	679	0	35	230	1	0	28	0	50
90	1	0	408	0	50	231	1	1	67	1	50
91	1	0	840	0	80	232	1	0	62	1	60
92	1	0	148	1	65	233	1	0	111	0	55
93	1	1	168	0	65	234	1	1	257	1	60
94	1	1	489	0	80	235	1	1	136	1	55

Table 11.1 Data in *heroin.dat*

95	1	0	541	0	80	236	1	0	342	0	60
96	1	1	205	0	50	237	2	1	41	0	40
97	1	0	475	1	75	238	2	0	531	1	45
98	1	1	237	0	45	239	1	0	98	0	40
99	1	1	517	0	70	240	1	1	145	1	55
100	1	1	749	0	70	241	1	1	50	0	50
101	1	1	150	1	80	242	1	0	53	0	50
102	1	1	465	0	65	243	1	0	103	1	50
103	2	1	708	1	60	244	1	0	2	1	60
104	2	0	713	0	50	245	1	1	157	1	60
105	2	0	146	0	50	246	1	1	75	1	55
106	2	1	450	0	55	247	1	1	19	1	40
109	2	0	555	0	80	248	1	1	35	0	60
110	2	1	460	0	50	249	2	0	394	1	80
111	2	0	53	1	60	250	1	1	117	0	40
113	2	1	122	1	60	251	1	1	175	1	60
114	2	1	35	1	40	252	1	1	180	1	60
118	2	0	532	0	70	253	1	1	314	0	70
119	2	0	684	0	65	254	1	0	480	0	50
120	2	0	769	1	70	255	1	0	325	1	60
121	2	0	591	0	70	256	2	1	280	0	90
122	2	0	769	1	40	257	1	1	204	0	50
123	2	0	609	1	100	258	2	1	366	0	55
124	2	0	932	1	80	259	2	0	531	1	50
125	2	0	932	1	80	260	1	1	59	1	45
126	2	0	587	0	110	261	1	1	33	1	60
127	2	1	26	0	40	262	2	1	540	0	80
128	2	0	72	1	40	263	2	0	551	0	65
129	2	0	641	0	70	264	1	1	90	0	40
131	2	0	367	0	70	266	1	1	47	0	45

The main reason why survival data require special methods of analysis is because they often contain right censored observations; that is, observations for which the endpoint of interest has not occurred during the period of observation; all that is known about the true survival time is that it exceeds the period of observation. Even if there are no censored observations, survival times tend to have positively skewed distributions that can be difficult to model.

11.2 Describing survival times and Cox's regression model

The survival time T can be regarded as a random variable with a probability distribution $F(t)$ and probability density function $f(t)$. Then an obvious function of interest is the probability of surviving to time t or beyond, the *survivor function* or survival curve $S(t)$, which is given by

$$S(t) = P(T \geq t) = 1 - F(t) \tag{11.1}$$

A further function of interest for survival data is the hazard function; this represents the instantaneous failure rate; that is, the probability that an individual expriences the event of interest at a time point given that the event has not yet occurred. It can be shown that the hazard function is given by

$$h(t) = \frac{f(t)}{S(t)} \tag{11.2}$$

the instananeous probability of failure at time t divided by the probability of surviving up to time t. Note that the hazard function is just the incidence rate discussed in Chapter 10. It follows from equations (11.1) and (11.2) that

$$\frac{-d\log(S(t))}{dt} = h(t) \qquad (11.3)$$

so that

$$S(t) = \exp(-H(t)) \qquad (11.4)$$

where $H(t)$ is the integrated hazard function, also known as the *cumulative hazard* function.

11.2.1 Cox's Regression

Cox's regression is a semiparametric approach to survival analysis. The method does not require the probability distribution $F(t)$ to be specified; however, unlike other nonparametric methods, Cox's regression does use regression parameters in the same way as generalized linear models. The model can be written as

$$h(t) = h_0(t)\exp(\boldsymbol{\beta}^T\mathbf{x}) \qquad (11.5)$$

so that the hazard functions of any two individuals are assumed to be constant multiples of each other, the multiple being $\exp(\boldsymbol{\beta}^T(\mathbf{x}_i - \mathbf{x}_j))$, the *hazard ratio* or incidence rate ratio. The assumption of a constant hazard ratio is called the *proportional hazards* assumption. The set of parameters $h_0(t)$, called the baseline hazard function, can be thought of as nuisance parameters whose purpose is merely to control the parameters of interest $\boldsymbol{\beta}$ for any changes in the hazard over time. It can be shown that the log profile likelihood (i.e., the log of the likelihood in which the nuisance parameters have been replaced by functions of $\boldsymbol{\beta}$ which maximize the likelihood for fixed $\boldsymbol{\beta}$) has the following form:

$$\sum_f \log\left(\frac{\exp\left(\boldsymbol{\beta}^T\mathbf{x}_f\right)}{\sum_{r(f)}\exp\left(\boldsymbol{\beta}^T\mathbf{x}_r\right)}\right) \qquad (11.6)$$

where the first summation is over all failures f and the second summation is over all subjects $r(f)$ still alive (and therefore "at risk") at the time of failure. The likelihood in equation (11.6) is equivalent to the likelihood for matched case-control studies described in Chapter 10 if the subjects at risk at the time of a failure (the *risk set*) are regarded as controls matched to the case failing at that point in time. However, despite this similarity, the likelihood is not a conditional likeklihood but a profile likelihood, also often referred to as a *partial likelihood* (see Clayton & Hills, 1993).

The baseline hazards can be estimated by maximising the partial likelihood with the regression parameters evaluated at their estimated values. These hazards are nonzero only when a failure occurs. Integrating the hazard function gives the cumulative hazard function

$$H(t) = H_0(t) \exp\left(\boldsymbol{\beta}^T \mathbf{x}\right) \tag{11.7}$$

where $H_0(t)$ is the integral of $h_0(t)$. The survival curve can be obtained from $H(t)$ using equation (11.4).

The log of the cumulative hazard function predicted by the Cox model is given by

$$\log(H(t)) = \log(H_0(t)) + \boldsymbol{\beta}^T \mathbf{x} \tag{11.8}$$

so that the log hazard functions of any two subjects i and j are parallel with constant difference given by $\boldsymbol{\beta}^T (\mathbf{x}_i - \mathbf{x}_j)$.

If the subjects fall into different groups and we are not sure whether we can make the assumption that the group's hazard functions are proportional to each other, we can estimate separate log cumulative hazard functions for the groups using a stratified Cox model. These curves can then be plotted to assess whether they are sufficiently parallel. For a stratified Cox model, the profile likelihood has the same form as in equation (11.6) except that the risk set for a failure is now confined to subjects in the same stratum.

Survival analysis is described in detail in Collett (1994) and in Clayton and Hills (1993) .

11.3 Analysis using Stata

The data are available as an ASCII file called *heroin.dat* on the disk accompanying Hand et al. (1994). Since the data are stored in a two-column format with the set of variables repeated twice in each row, as shown in Table 11.1, we have to use reshape to bring the data into the usual form:

```
infile id1 clinic1 status1 time1 prison1 dose1 /*
    */ id2 clinic2 status2 time2 prison2 dose2 using heroin.dat
gen row=_n
reshape groups col 1-2
reshape cons row
reshape vars id clinic status time prison dose
reshape long
drop row col
```

Before fitting any survival time models, we declare the data as being of the form "st" (for survival time)

```
stset time status
```

and look at a summary of the data using

```
stsum
```

```
 failure time:   time
 failure/censor:  status

         |                incidence      no. of    |------ Survival time -----|
         | time at risk      rate       subjects       25%      50%       75%
---------+----------------------------------------------------------------------
 total   |        95812    .0015656          238       212      504       821
```

There are therefore 238 subjects with a median survival time of 504 days. If the incidence rate (i.e., the hazard function) could be assumed to be constant, it would be estimated as 0.0016 per day, which corresponds to 0.57 per year.

The data come from two different clinics and it is likely that these clinics have different hazard functions which may well not be parallel. A Cox regression model with clinics as strata and the other two variables, *dose* and *prison*, as explanatory variables is fitted using

```
stcox dose prison, strata(clinic)
```

```
   failure time:   time
   failure/censor:   status
Iteration 0:   Log Likelihood =-614.68365
Iteration 1:   Log Likelihood =-597.73516
Iteration 2:   Log Likelihood = -597.714
Refining estimates:
Iteration 0:   Log Likelihood = -597.714
Stratified Cox regr. -- entry time 0
No. of subjects =         238              Log likelihood =    -597.714
No. of failures =         150              chi2(2)        =      33.94
Time at risk    =       95812              Prob > chi2    =     0.0000
------------------------------------------------------------------------------
    time |
  status | Haz. Ratio   Std. Err.       z      P>|z|     [95% Conf. Interval]
---------+--------------------------------------------------------------------
    dose |   .9654655   .0062418     -5.436    0.000      .953309     0.977777
  prison |   1.475192   .2491827      2.302    0.021     1.059418     2.054138
------------------------------------------------------------------------------
                                                        Stratified by clinic
```

Therefore, subjects with a prison history are 47.5% more likely to drop out at any given time than those without a prison history. For every increase in methadone dose by one unit (1mg), the hazard is multiplied by 0.965. This coefficient is very close to one, but this may be because one unit of methadone dose is not a large quantity. Even if we know little about methadone mainte- nance treatment, we can assess how much one unit of methadone dose is by finding the sample standard deviation:

```
summarize dose
```

```
Variable |      Obs        Mean    Std. Dev.        Min         Max
---------+----------------------------------------------------------
    dose |      238    60.39916    14.45013          20         110
```

indicating that a unit is not much at all; subjects often differ from each other by 10 to 15 units. In order to find the hazard ratio of two subjects differing by one standard deviation, we need to raise the hazard ratio to the power of one standard deviation, giving $0.9654655^{14.45013} = 0.60179167$. We can obtain the same (but more precise) result using the stored macros _b[dose] for the log hazard ratio and _result(4) for the variance:

```
display exp(_b[dose]*sqrt(_result(4)))
```

$$\boxed{.60178874}$$

In the above calculation, we simply rescaled the regression coefficient before taking the exponential. This can also be achieved by rescaling the variable itself, i.e., by standardizing *dose* to have unit standard deviation. In the command below, we also standardize to mean zero although this will make no difference to the estimated coefficients (except the constant):

```
replace dose=(dose-_result(3))/sqrt(_result(4))
```

we repeat the Cox regresion with the option bases(s), which results in the survival function $S_0(t)$ being estimated and stored in s:

```
stcox dose prison, strata(clinic) bases(s)
```

```
No. of subjects =        238              Log likelihood =    -597.714
No. of failures =        150              chi2(2)        =       33.94
Time at risk    =      95812              Prob > chi2    =      0.0000

------------------------------------------------------------------------
    time |
  status | Haz. Ratio   Std. Err.      z    P>|z|    [95% Conf. Interval]
---------+--------------------------------------------------------------
    dose |   .6017887    .0562195   -5.436   0.000    .5010998    .7227097
  prison |   1.475192    .2491827    2.302   0.021    1.059418    2.054138
------------------------------------------------------------------------
                                              Stratified by clinic
```

The coefficient of *dose* is identical to that calculated previously and can now be interpreted as indicating a decrease of the hazard by 40% when the methadone dose increases by one standard deviation.

One question to investigate is whether the clinics need to be treated as strata or whether the hazard functions are in fact proportional. We can do this by plotting the log cumulative hazard function for the two clinics and visually assessing whether these curves are approximately parallel. The variables required for the graph are obtained as follows:

```
gen lh1=log(-log(s)) if clinic==1
gen lh2=log(-log(s)) if clinic==2
sort clinic time
by clinic: list clinic status time lh1 lh2 if _n==_N
```

```
-> clinic=          1
        clinic      status      time        lh1         lh2
163.         1           0        905   1.477235           .


-> clinic=          2
        clinic      status      time        lh1         lh2
238.         2           0       1076           .   -.4802778
```

The curves therefore end at times 905 and 1075, respectively, with corresponding
values on the vertical axis of 1.48 and −0.48. This information can be used to
define the positions for labels "clin1" and "clin2" for the curves as follows:

```
gen pos1=1.48+0.2 if time==905
gen str5 lab1="clin1"
gen pos2=-0.48+0.2 if time==1076
gen str5 lab2="clin2"
```

The graph is produced using

```
graph lh1 lh2 pos1 pos2 time, sort s(.i[lab1][lab2]) c(JJ..) /*
    */ xlabel ylabel pen(6311) psize(130) t1("     ")         /*
    */ l1("Log of cumulative hazard") gap(3)
```

Here, the option connect(J) causes points to be connected using a step function
and pen(6311) was used to control the colors (and thicknesses) of the lines and
plotting symbols, mainly to have the labels appear in the same color (pen=1) as
the axes and titles. The resulting graph is shown in Figure 11.1. In these curves,
the increment at each failure represents the estimated log of the hazards at that
time. Clearly, the curves are not parallel and we will therefore continue treating
the clinics as strata.

11.3.1 Model presentation and diagnostics

Assuming the variables *prison* and *dose* satisfy the proportional hazards assumption (see Section 11.3.2), a discussion of how to present the model follows.

A good graph for presenting the results of a Cox regression is a graph of the
survival curves fitted for groups of subjects when the continuous variables take
on their mean values. Such a graph can be produced by first generating variable
surv1 and *surv2* to hold the fitted survival curves for the two groups.

```
stcox dose prison, strata(clinic) bases(s_strat)
egen mdose=mean(dose), by(clinic)
gen surv1=s_strat^exp(_b[dose]*mdose) if prison==0
gen surv2=s_strat^exp(_b[dose]*mdose+_b[prison]) if prison==1
```

Note that the survival functions represent the expected survival function for a
subject having the clinic-specific mean dose. We now transform *time* to time in
years and plot the survival curves separately for each clinic:

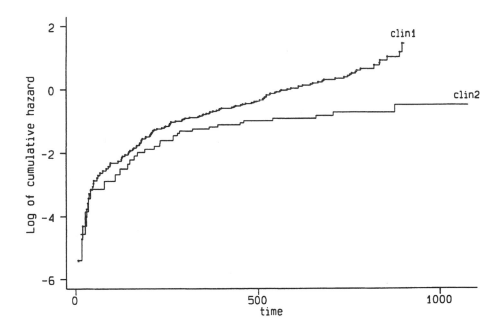

Figure 11.1 *Log of minus the log of the survival functions for the two clinics estimated by stratified Cox regression.*

```
gen tt=time/365.25
label variable tt "time in years"
graph surv1 surv2 tt, sort c(JJ) s(..) by(clinic) t1(" ") /*
      */ xlabel ylabel(0.0, 0.2, 0.4, 0.6, 0.8, 1.0)      /*
```
resulting in two graphs displayed side by side at the top of the graphics window separated by a wide gap from a title at the bottom of the window. A better-looking figure can be produced by displaying separate graphs for the clinic side by side, where the separate graphs are created using

```
graph surv1 surv2 tt if clinic==1 , sort c(JJ) s(..)          /*
      */ xlabel(0,1,2,3) ylabel(0.0, 0.2, 0.4, 0.6, 0.8, 1.0) /*
      */ t1("clinic 1") l1("fraction remaining") gap(3)
```
and similarly for clinic 2. The result is shown in Figure 11.2.

According to Caplehorn and Bell (1991), the more rapid decline in the proportion remaining in clinic 1 compared with clinic 2 may be due to the policy of clinic 1 to attempt to limit the duration of maintenance to 2 years.

It is probably a good idea to plot some residuals, for example, the deviance residuals. The deviance residuals can be computed from the martingale residuals (see, for example, Collett, 1994), which are obtained using the option `mgale` as follows:

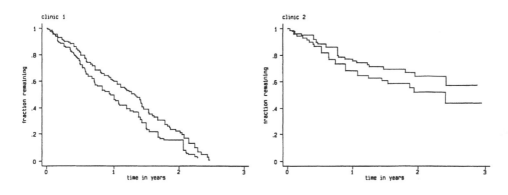

Figure 11.2 *Survival curves.*

```
stcox dose prison, strata(clinic) mgale(mart)
gen devr=sign(mart)*sqrt(-2*(mart+status*log(status-mart)))
predict xb, xb
graph devr xb, xlabel ylabel s([id]) ll(''Deviance Residuals'')
gap(3)
```

with the result shown in Figure 11.3. There appear to be no serious outliers.

Another type of residual is a score residual, defined as the first derivative of the log partial likelihood function with respect to an explanatory variable. The score residual is large in absolute value if a case's explanatory variable differs substantially from the the explanatory variables of subjects whose estimated risk of failure is large at the case's time of failure or censoring. Since this model has two explanatory variables, we can derive two score residuals, *score1* and *score2*, using the command

```
stcox dose prison, strata(clinic) bases(s_strat) esr(score*)
```

These residuals represent score residuals for *clinic* and *dose*, respectively, and can be plotted against survival time using

```
label variable score1 "score residuals for dose"
graph score1 tt, s([id]) gap(3) xlabel ylabel
```

and similarly for *score2*. The resulting graphs are shown in Figures 11.4 and 11.5. Subject 89 has a low dose (−1.75) compared with other subjects at risk of failure at such a late time. Subjects 8, 27, 12, and 71 fail relatively late considering that they have a police record, whereas other remainers tend to have no police record at their time of dropping out (or censoring).

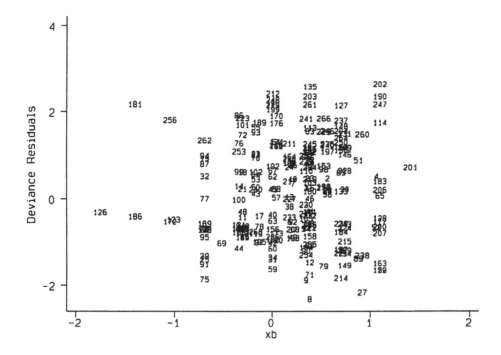

Figure 11.3 *Deviance residuals for survival analysis.*

11.3.2 *Cox's regression with time-varying covariates*

In order to determine whether the hazard functions for those with and without a prison history are proportional, we could split the data into four strata by *clinic* and *prison*. However, as the strata get smaller, the estimated survival functions become less precise (due to the risk sets in equation (11.6) becoming smaller). Also, a similar method could not be used to check the proportional hazard assumption for the continuous variable *dose*. Another way of testing the proportional hazards assumption of *dose*, say, is to introduce a time-varying covariate equal to the interaction between the survival time variable and *dose*. In order to estimate this model, the terms in equation (11.6) need to be evaluated using the values of the time-varying covariates *at the times of the failure*. These values are not available in the present dataset since each subject is represented only once, at the time of their own failure. We therefore need to expand the dataset so that each subject's record appears (at least) as many times as that subject contributes to a risk set in equation (11.6), with the time variable equal to the corresponding failure times. The simplest way to achieve this for discrete survival times is to represent each subject by *time* observations with *time* equal to 1,2,···,*time* and to define a new status variable *fail*, which is zero for all except

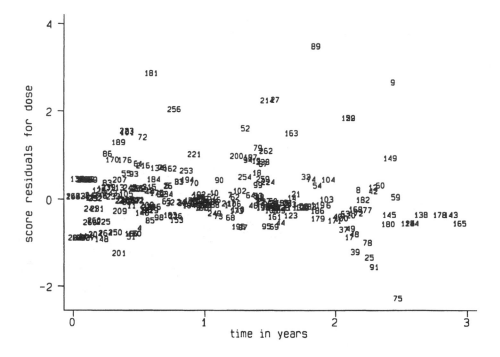

Figure 11.4 *Score residuals for prison.*

the last observation for each subject which is equal to *status*. The following code
is adapted from the Stata FAQ by W. Gould (1997) .

```
expand time
sort id
quietly by id: gen t = _n
gen fail = 0
quietly by id: replace fail = status if _n==_N
```

A problem with applying this code to this data is that the survival times are
measured in days, with a median survival of 504 days. Unless we are willing
to round time, for example, to integer number of months, we would have to
replicate the data by a factor of about 500! Therefore, a more feasible method
is described.

Before beginning, we must make sure that there is enough memory allocated
to Stata to hold the increased dataset. The memory can be set to 3 megabytes
using

```
clear
set memory 3m
```

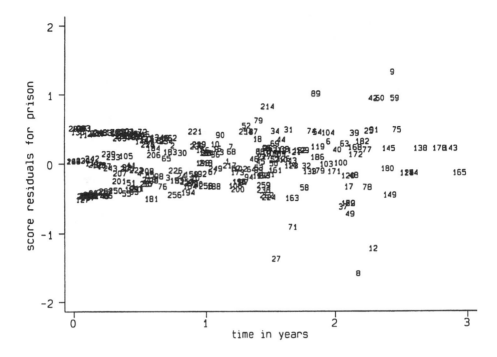

Figure 11.5 *Score residuals for dose.*

(the data have to be read in again after this). If `set memory` does not work, add the /k3000 option to the command for running Stata, e.g.,

```
C:\Stata\wstata.exe /k3000
```

in the "Target" field of the "Properties (Shortcut)" dialog box for Stata. We also need to define matrices and in order to ensure that we do not exceed the maximum matrix size, the limit is increased to 300 using

```
set matsize 300
```

The first step is to determine the sets of unique failure times for the two clinics and store these in matrices `t1` and `t2` using the command `mkmat` to place observations from the dataset into a matrix:

```
sort clinic status time
gen unft=0 /* indicator for unique failure times */
quietly by clinic status time: replace unft=1 if _n==1&status==1

sort time
mkmat time if (unft==1)&clinic==1,matrix(t1)
sort time
```

```
mkmat time if (unft==1)&clinic==2,matrix(t2)
```
We can look at the contents of t2 using

```
matrix list t2
```

with the result shown in Display 11.1. Note that t2 is a matrix although it only

```
t2[27,1]
          time
   r1       13
   r2       26
   r3       35
   r4       41
   r5       79
   r6      109
   r7      122
   r8      143
   r9      149
  r10      161
  r11      170
  r12      190
  r13      216
  r14      231
  r15      232
  r16      268
  r17      280
  r18      286
  r19      322
  r20      366
  r21      389
  r22      450
  r23      460
  r24      540
  r25      661
  r26      708
  r27      878
```

Display 11.1

holds a one-dimensional array of numbers. This is because there are no vectors in
Stata. The matrix is $r \times 1$ where r can be found using the function rowsof(t2).
Elements of the matrix can be accessed using the expression t2[3,1] for the
third unique failure time in clinic2.

Next, determine *num*, the number of replicates required of each subject's
record. Initially, *num* is set equal to the number of unique survival times in
each subject's clinic, i.e., the number of rows of the corresponding matrix:

```
local nt1=rowsof(t1)
local nt2=rowsof(t2)
gen num=cond(clinic==1,'nt1','nt2')
```

The number of required replicates is however just the number of unique survival times *which are less than* the subject's own survival time, *time*, because

the subject ceases to contribute to future risk sets once it has failed. Therefore, reduce *num* until t1[num,1] is less than *time* for the first time, implying that t1[num+1,1] is greater than *time*:

```
quietly for 1-'nt1', ltype(numeric): replace num=num-1   /*
     */ if clinic==1&time<=t1['nt1'-@+1,1]
quietly for 1-'nt2', ltype(numeric): replace num=num-1   /*
     */ if clinic==2&time<=t2['nt2'-@+1,1]
```

The commands above take a long time to execute and time could be saved by writing a program and making use of while (see exercises). One extra record is needed for each subject to hold their own survival time:

```
replace num=num+1
```

We are now ready to expand the dataset and fill in the unique survival times for the first *num*-1 records of each subject, followed by the subject's own survival time in the last record. All except the last values of *failure* are 0 (for censored) and the last value is equal to the subject's status at their own survival time:

```
compress
expand num

sort id time
quietly by id: gen t=cond(clinic==1,t1[_n,1],t2[_n,1])
quietly by id: replace t=time if _n==_N

gen fail=0 /* indicator for failure */
sort id t
quietly by id: replace fail=status if _n==_N
```

The command compress was used before expand to ensure that the variables are stored as compactly as possible before another 10990 observations were created. The data are now in a suitable form to perfom Cox regression with time-varying covariates. First verify that the data are equivalent to the original form by repeating the simple Cox regression

```
stset t fail, id(id)
stcox dose prison, strata(clinic)
```

which gives the same result as before. Now generate a time-varying covariate for *dose* as follows:

```
gen tdose=dose*(t-504)/365.25
```

where we have subtracted the median survival time so that the effect of the explanatory variable *dose* can be interpreted as the effect of *dose* at the median survival time. Divide by 365.25 in order to see by how much the effect of *dose* changes between intervals of one year. Now fit the Cox regression, allowing the effect of *dose* to vary with time.

```
stcox dose tdose prison, strata(clinic)
```

```
        failure time:  t
          entry time:  t0
      failure/censor:  fail
                  id:  id

Iteration 0:  Log Likelihood =-614.68365
Iteration 1:  Log Likelihood =-597.32655
Iteration 2:  Log Likelihood =-597.29131
Iteration 3:  Log Likelihood =-597.29131
Refining estimates:
Iteration 0:  Log Likelihood =-597.29131

Stratified Cox regr. -- entry time t0

No. of subjects =         238            Log likelihood =  -597.29131
No. of failures =         150            chi2(3)        =       34.78
Time at risk    =       95812            Prob > chi2    =      0.0000

-----------------------------------------------------------------------
    t |
 fail | Haz. Ratio  Std. Err.      z    P>|z|    [95% Conf. Interval]
------+----------------------------------------------------------------
 dose |  .6442974    .0767535   -3.690  0.000    .5101348    .8137442
tdose | 1.147853     .1720104    0.920  0.357    .8557175   1.539722
prison| 1.481193     .249978     2.328  0.020   1.064036    2.061899
-----------------------------------------------------------------------
                                                    Stratified by clinic
```

Display 11.2

In the output shown in Display 11.2, the estimated increase in the hazard ratio for a one standard deviation increase in *dose* is 15% for each increase in time by one year. This small effect is not significant at the 5% level, which is confirmed by carrying out the likelihood ratio test as follows:

```
lrtest, saving(0)
stcox dose prison, strata(clinic)
lrtest
```

```
Cox: likelihood-ratio test                chi2(1)    =      0.85
                                          Prob > chi2 =    0.3579
```

giving a very similar *p*-value as before and confirming that there is no evidence that the effect of dose on the hazard varies with time. A similar test can be carried out for *prison* (see exercise 5.).

We can now return to the original data, either by reading them again or by dropping all the newly created observations as follows:

```
sort id t
quietly by id: drop if _n<_N
stset t fail
stcox dose prison, strata(clinic) bases(s_strat)
```

11.4 Exercises

1. In the original analysis of this data, Caplehorn and Bell (1991) judged that the hazards in the two clinics were approximately proportional for the first 450 days (see Figure 11.1). They therefore analyzed the data for this time period using *clinic* as a covariate instead of stratifying by clinic. Repeat this analysis, using *prison* and *dose* as further covariates.

2. Following Caplehorn and Bell (1991), repeat the above analysis treating dose as a categorical variable with three levels (< 60, 60–79, ≥ 80) and plot the predicted survival curves for the three dose categories when *prison* and *clinic* take on one of their values.

3. Test for an interaction between the clinic and the methadone dose using both continuous and categorical scales for dose.

4. Create a "do-file" containing the commands given above to expand the data for Cox regression with time-varying covariates.

5. Read and expand the data (using the "do-file") and check the proportional hazards assumption for *prison* using the same method used for *dose*.

6. The `for` command is very slow; speed up the reduction of *num* by replacing the `for` commands by a small program that uses `while`.

CHAPTER 12

Principal Components Analysis: Hearing Measurement Using an Audiometer

12.1 Description of data

The data in Table 12.1 are adapted from those given in Jackson (1991), and relate to hearing measurement with an instrument called an audiometer. An individual is exposed to a signal of a given frequency with an increasing intensity until the signal is perceived. The lowest intensity at which the signal is perceived is a measure of hearing loss, calibrated in units referred to as *decibel loss* in comparison to a reference standard for that particular instrument. Observations are obtained one ear at a time, for a number of frequencies. In this example, the frequencies used were 500 Hz, 1000 Hz, 2000 Hz, and 4000 Hz. The limits of the instrument are -10 to 99 decibels. (A negative value does not imply better than average hearing; the audiometer had a calibration 'zero' and these observations are in relation to that.)

Table 12.1 Data in *hear.dat*
(Taken from Jackson (1992) with permission of his publisher, John Wiley and Sons)

id	l500	l1000	l2000	l4000	r500	r1000	r2000	r4000
1	0	5	10	15	0	5	5	15
2	−5	0	−10	0	0	5	5	15
3	−5	0	15	15	0	0	5	15
4	−5	0	−10	−10	−10	−5	−10	10
5	−5	−5	−10	10	0	−10	−10	50
6	5	5	5	−10	0	5	0	20
7	0	0	0	20	5	5	5	10
8	−10	−10	−10	−5	−10	−5	0	5
9	0	0	0	40	0	0	−10	10
10	−5	−5	−10	20	−10	−5	−10	15
11	−10	−5	−5	5	5	0	−10	5
12	5	5	10	25	−5	−5	5	15
13	0	0	−10	15	−10	−10	−10	10

Table 12.1 Data in *hear.dat*
(Taken from Jackson (1992) with permission of his publisher, John Wiley and Sons)

14	5	15	5	60	5	5	0	50
15	5	0	5	15	5	−5	0	25
16	−5	−5	5	30	5	5	5	25
17	0	−10	0	20	0	−10	−10	25
18	5	0	0	50	10	10	5	65
19	−10	0	0	15	−10	−5	5	15
20	−10	−10	−5	0	−10	−5	−5	5
21	−5	−5	−5	35	−5	−5	−10	20
22	5	15	5	20	5	5	5	25
23	−10	−10	−10	25	−5	−10	−10	25
24	−10	0	5	15	−10	−5	5	20
25	0	0	0	20	−5	−5	10	30
26	−10	−5	0	15	0	0	0	10
27	0	0	5	50	5	0	5	40
28	−5	−5	−5	55	−5	5	10	70
29	0	15	0	20	10	−5	0	10
30	−10	−5	0	15	−5	0	10	20
31	−10	−10	5	10	0	0	20	10
32	−5	5	10	25	−5	0	5	10
33	0	5	0	10	−10	0	0	0
34	−10	−10	−10	45	−10	−10	5	45
35	−5	10	20	45	−5	10	35	60
36	−5	−5	−5	30	−5	0	10	40
37	−10	−5	−5	45	−10	−5	−5	50
38	5	5	5	25	−5	−5	5	40
39	−10	−10	−10	0	−10	−10	−10	5
40	10	20	15	10	25	20	10	20
41	−10	−10	−10	20	−10	−10	0	5
42	5	5	−5	40	5	10	0	45
43	−10	0	10	20	−10	0	15	10
44	−10	−10	10	10	−10	−10	5	0
45	−5	−5	−10	35	−5	0	−10	55
46	5	5	10	25	10	5	5	20
47	5	0	10	70	−5	5	15	40
48	5	10	0	15	5	10	0	30
49	−5	−5	5	−10	−10	−5	0	20
50	−5	0	10	55	−10	0	5	50
51	−10	−10	−10	5	−10	−10	−5	0
52	5	10	20	25	0	5	15	0

Table 12.1 Data in *hear.dat*
(Taken from Jackson (1992) with permission of his publisher, John Wiley and Sons)

53	−10	−10	50	25	−10	−10	−10	40
54	5	10	0	−10	0	5	−5	15
55	15	20	10	60	20	20	0	25
56	−10	−10	−10	5	−10	−10	−5	−10
57	−5	−5	−10	30	0	−5	−10	15
58	−5	−5	0	5	−5	−5	0	10
59	−5	5	5	40	0	0	0	10
60	5	10	30	20	5	5	20	60
61	5	5	0	10	−5	5	0	10
62	0	5	10	35	0	0	5	20
63	−10	−10	−10	0	−5	0	−5	0
64	−10	−5	−5	20	−10	−10	−5	5
65	5	10	0	25	5	5	0	15
66	−10	0	5	60	−10	−5	0	65
67	5	10	40	55	0	5	30	40
68	−5	−10	−10	20	−5	−10	−10	15
69	−5	−5	−5	20	−5	0	0	0
70	−5	−5	−5	5	−5	0	0	5
71	0	10	40	60	−5	0	25	50
72	−5	−5	−5	−5	−5	−5	−5	5
73	0	5	45	50	0	10	15	50
74	−5	−5	10	25	−10	−5	25	60
75	0	−10	0	60	15	0	5	50
76	−5	0	10	35	−10	0	0	15
77	5	0	0	15	0	5	5	25
78	15	15	5	35	10	15	−5	0
79	−10	−10	−10	5	−5	−5	−5	5
80	−10	−10	−5	15	−10	−10	−5	5
81	0	−5	5	35	−5	−5	5	15
82	−5	−5	−5	10	−5	−5	−5	5
83	−5	−5	−10	−10	0	−5	−10	0
84	5	10	10	20	−5	0	0	10
85	−10	−10	−10	5	−10	−5	−10	20
86	5	5	10	0	0	5	5	5
87	−10	0	−5	−10	−10	0	0	−10
88	−10	−10	10	15	0	0	5	15
89	−5	0	10	25	−5	0	5	10
90	5	0	−10	−10	10	0	0	0
91	0	0	5	15	5	0	0	5

Table 12.1 Data in *hear.dat*
(Taken from Jackson (1992) with permission of his publisher, John
Wiley and Sons)

92	−5	0	−5	0	−5	−5	−10	0
93	−5	5	−10	45	−5	0	−5	25
94	−10	−5	0	10	−10	5	−10	10
95	−10	−5	0	5	−10	−5	−5	5
96	5	0	5	0	5	0	5	15
97	−10	−10	5	40	−10	−5	−10	5
98	10	10	15	55	0	0	5	75
99	−5	5	5	20	−5	5	5	40
100	−5	−5	−10	−10	−5	0	15	10

12.2 Principal components analysis

Principal components analysis is one of the oldest but still most widely used techniques of multivariate analysis. Originally introduced by Pearson (1901) and independently by Hotelling (1933), the basic idea of the method is to try to describe the variation of the variables in a set of multivariate data as parsimoniously as possible using a set of derived uncorrelated variables, each of which is a particular linear combination of those in the original data. In other words, principal components analysis is a transformation from the observed variables, $x_1, \cdots, x_p$ to new variables $y_1, \cdots, y_p$ where

$$
\begin{aligned}
y_1 &= a_{11}x_1 + a_{12}x_2 + \cdots + a_{ip}x_p \\
y_2 &= a_{21}x_1 + a_{22}x_2 + \cdots + a_{2p}x_p \\
\vdots &= \vdots + \vdots + \vdots + \vdots \\
y_p &= a_{p1}x_1 + a_{p2}x_2 + \cdots + a_{pp}x_p
\end{aligned}
\tag{12.1}
$$

The new variables are derived in decreasing order of importance so that the first principal component variable (y_1) accounts for as much of the variation of the original variables as possible. The second principal component y_2 accounts for as much of the remaining variation as possible *conditional* on being uncorrelated with y_1 and so on. The usual objective of this type of analysis is to assess whether the first few components account for a substantial proportion of the variation in the data. If they do, they can be used to summarize the data with little loss of information. This can be useful for obtaining graphical displays of the multivariate data or for simplifying subsequent analysis.

The coefficients defining the principal components are obtained from the eigenvalues of either the covariance or correlation matrix of the original variables (giving different results). The variances of the derived variables are given by the eigenvalues of the corresponding matrix. A detailed account of principal components analysis is given in Everitt and Dunn (1991).

12.3 Analysis using Stata

The data can be read in from an ASCII file *hear.dat* as follows:

```
infile id l500 l1000 l2000 l4000 r500 r1000 r2000 r4000 using
   hear.dat
summarize
```

Variable	Obs	Mean	Std. Dev.	Min	Max
id	100	50.5	29.01149	1	100
l500	100	-2.8	6.408643	-10	15
l1000	100	-.5	7.571211	-10	20
l2000	100	2.45	11.94463	-10	50
l4000	100	21.35	19.61569	-10	70
r500	100	-2.6	7.123726	-10	25
r1000	100	-.7	6.396811	-10	20
r2000	100	1.55	9.257675	-10	35
r4000	100	20.95	19.43254	-10	75

Before undertaking a principal components analysis, some graphical exploration of the data might be useful. A scatter-plot matrix, for example, with points labeled with a subject's identification number can be obtained using

```
graph l500-r4000, matrix half symbol([id]) ps(150)
```

The resulting diagram is shown in Figure 12.1. The diagram looks a little 'odd' due to the largely 'discrete' nature of the observations. Some of the individual scatter-plots suggest that some of the observations might perhaps be regarded as outliers; for example, individual 53 in the plot involving *l2000, r2000*. This subject has a score of 50 at this frequency in the left ear, but a score of -10 in the right ear. It might be appropriate to remove this subject's observations before further analysis, but we do not do this and continue to use the data from all 100 individuals.

As mentioned in the previous section, principal components can be extracted from either the covariance matrix or the correlation matrix of the original variables. A choice needs to be made since there is not necessarily any simple relationship between the results in each case. The summary table shows that the variances of the observations at the highest frequencies are approximately nine times those at the lower frequencies; consequently, a principal components analysis using the covariance matrix would be dominated by the 4000-Hz frequency. But this frequency is not more clinically important that the others, and so, in this case, it seems more reasonable to use the correlation matrix as the basis of the principal components analysis.

To find the correlation matrix of the data requires the following instruction:

```
correlate l500-r4000
```

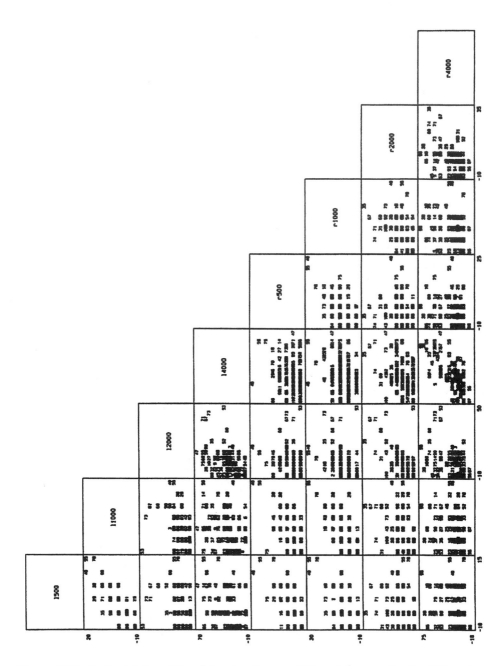

Figure 12.1 *Scatter-plot matrix of hearing loss at different frequencies for left and right ear.*

```
(obs=100)
        |    1500    11000    12000    14000     r500    r1000    r2000
--------+-------------------------------------------------------------------
  1500|  1.0000
 11000|  0.7775   1.0000
 12000|  0.3247   0.4437   1.0000
 14000|  0.2554   0.2749   0.3964   1.0000
  r500|  0.6963   0.5515   0.1795   0.1790   1.0000
 r1000|  0.6416   0.7070   0.3532   0.2632   0.6634   1.0000
 r2000|  0.2399   0.3606   0.5910   0.3193   0.1575   0.4151   1.0000
 r4000|  0.2264   0.2109   0.3598   0.6783   0.1421   0.2248   0.4044
        |   r4000
--------+----------
 r4000|  1.0000
```

Note that the highest correlations occur between adjacent frequencies on the
same ear and between corresponding frequencies on different ears.

To obtain the principal components of this correlation matrix requires the
use of the factor procedure:

```
factor 1500-r4000, pc
```

which gives the results shown in Display 12.1. The pc option is used to obtain
principal components rather than a factor analysis solution. An informal rule for
choosing the number of components to represent a set of correlations is to use
only those components with eigenvalues greater than 1. Here, this would lead
to retaining only the first two components. Another informal indicator of the
appropriate number of components is the *scree-plot*, a plot of the eigenvalues.
A scree-plot can be obtained using

```
greigen, gap(3)
```

with the result shown in Figure 12.2. An 'elbow' in the scree-plot indicates the
number of eigenvalues to choose. From Figure 9.2 this would again appear to
be 2. The first two components account for 68% of the variance in the data.

The elements of the eigenvalues defining the principal components are scaled
so that their sums of squares are unity. A more useful scaling is often obtained
from multiplying the elements by the square root of the corresponding eigen-
value, in which case the coefficients represent correlations between an observed
variable and a component.

Examining the eigenvectors defining the first two principal components, we
see that the first accounting for 48% of the variance has coefficients that are all
positive and all approximately the same size. This principal component represents, essentially, the overall hearing loss of a subject and implies that individuals suffering hearing loss at certain frequencies will be more likely to suffer this
loss at other frequencies as well. The second component, accounting for 20% of
the variance contrasts high frequencies (2000 Hz and 4000 Hz) and low frequencies (500 Hz and 1000 Hz). It is well known in the case of normal hearing that
hearing loss as a function of age is first noticeable in the higher frequencies.

```
(obs=100)

                (principal components; 8 components retained)
Component     Eigenvalue     Difference     Proportion     Cumulative
-----------------------------------------------------------------------
    1          3.82375        2.18915         0.4780         0.4780
    2          1.63459        0.72555         0.2043         0.6823
    3          0.90904        0.40953         0.1136         0.7959
    4          0.49951        0.12208         0.0624         0.8584
    5          0.37743        0.03833         0.0472         0.9055
    6          0.33910        0.07809         0.0424         0.9479
    7          0.26101        0.10545         0.0326         0.9806
    8          0.15556           .            0.0194         1.0000

             Eigenvectors
Variable |      1          2          3          4          5          6
---------+--------------------------------------------------------------
   1500 |   0.40915    -0.31257    0.13593    -0.27217    -0.16650    0.41679
  11000 |   0.42415    -0.23011   -0.09332    -0.35284    -0.49977   -0.08474
  12000 |   0.32707     0.30065   -0.47772    -0.48723     0.50331    0.04038
  14000 |   0.28495     0.44875    0.47110    -0.17955     0.09901   -0.51286
   r500 |   0.35112    -0.38744    0.23944     0.30453     0.62830    0.17764
  r1000 |   0.41602    -0.23673   -0.05684     0.36453    -0.08611   -0.54457
  r2000 |   0.30896     0.32280   -0.53841     0.51686    -0.16229    0.12553
  r4000 |   0.26964     0.49723    0.41499     0.19757    -0.17570    0.45889

             Eigenvectors
Variable |      7          8
---------+---------------------
   1500 |   0.28281    -0.60077
  11000 |  -0.02919     0.61330
  12000 |  -0.27925    -0.06396
  14000 |   0.43536    -0.02978
   r500 |   0.12745     0.36603
  r1000 |  -0.46180    -0.34285
  r2000 |   0.44761     0.02927
  r4000 |  -0.47094     0.07469
```

Display 12.1

Scores for each individual on the first two principal components might be used as a convenient way of summarizing the original eight dimensional data. Such scores are obtained by applying the elements of the corresponding eigenvector to the standardized values of the original observations for an individual. The necessary calculations can be carried out with the score procedure:

```
score pc1 pc2
```

```
graph pc2 pc1, symbol([id]) ps(150)
```

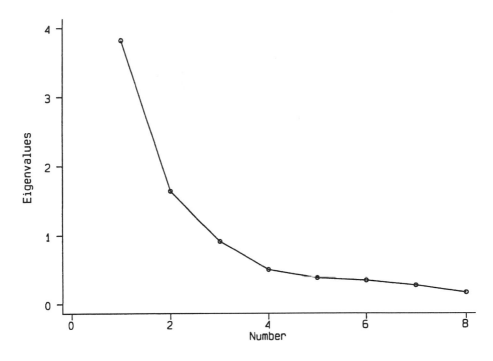

Figure 12.2 *Scree-plot.*

	(based on unrotated principal components) (6 scorings not used)	
	Scoring Coefficients	
Variable |	1	2
1500 |	0.40915	-0.31257
11000 |	0.42415	-0.23011
12000 |	0.32707	0.30065
14000 |	0.28495	0.44875
r500 |	0.35112	-0.38744
r1000 |	0.41602	-0.23673
r2000 |	0.30896	0.32280
r4000 |	0.26964	0.49723

The new variables *pc1* and *pc2* contain the scores for the first two principal components and the output lists the coefficients used to form these scores. For principal components analysis, these coefficients are just the eigenvalues in Display 12.1. The principal component scores can be used to produce a useful graphical display of the data in a single scatter-plot, which can then be used to search for structure or patterns in the data, particularly the presence of clusters of observations (see Everitt, 1993). Note that the distances between observations in this graph approximate the Euclidean distances between the (standardized) variables, i.e., the graph is a *multidimensional scaling* solution. In fact, the graph is the classical scaling (or principal coordinate) scaling solution to the Euclidean distances (see Everitt and Dunn, 1991, or Everitt and Rabe-Hesketh, 1997).

The principal component plot is obtained using

```
graph pc2 pc1, symbol([id]) ps(150) gap(3) xlab ylab
```

The resulting diagram is shown in Figure 12.3. Here, the variablility in differential hearing loss for high versus low frequencies (*pc2*) is greater among subjects with higher overall hearing loss, as would be expected. It would be interesting to investigate the relationship between the principal components and other variables related to hearing loss such as age (see exercise 6).

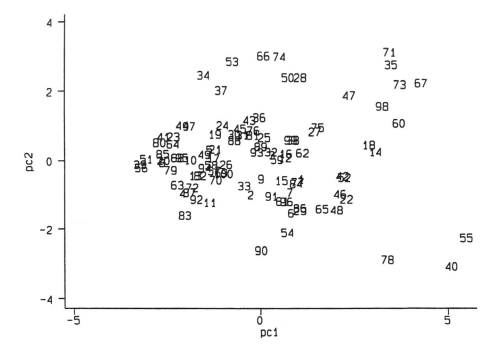

Figure 12.3 *Principal component plot.*

12.4 Exercises

1. Rerun the principal components analysis described in this chapter using the covariances matrix of the observations. Compare the results with those based on the correlation matrix.

2. Interpret components 3–8 in the principal components analysis based on the correlation matrix.

3. Create a scatter-plot matrix of the first five principal components scores.

4. Investigate other methods of factor analysis available in `factor` applied to the hearing data.

5. Apply principal component analysis to the air pollution data analyzed in Chapter 3, exluding the variable *so2*, and plot the first two principal components (i.e., the 2-d classical scaling solution for Euclidean distances between standardized variables)

6. Regress *so2* on the first two principal components and add a line corresponding to this regression (the direction of steepest increase in *so2* predicted by the regression plane) into the multidimensional scaling solution.

CHAPTER 13

Maximum Likelihood Estimation: Age of Onset of Schizophrenia

13.1 Description of data

In Table 13.1, the ages of onset of schizophrenia in 99 women (determined as age on first admission) are given. These data will be used in order to investigate whether there is any evidence for the subtype model of schizophrenia (see Lewine, 1981). According to this model, there are two types of schizophrenia. One is characterized by early onset, typical symptoms, and poor premorbid competence and the other by late onset, atypical symptoms, and good premorbid competence. We will investigate this question by fitting a mixture of two normal distribution to the ages.

13.2 Finite mixture distributions

The most common type of finite mixture distribution encountered in practice is a mixture of univariate Gaussian distributions of the form

$$f\left(y_i; \mathbf{p}, \boldsymbol{\mu}, \boldsymbol{\sigma}\right) = p_1 g\left(y_i; \mu_1, \sigma_1\right) + p_2 g\left(y_i; \mu_1, \sigma_1\right) + \cdots + p_k g\left(y_i; \mu_1, \sigma_1\right) \quad (13.1)$$

where $g(y; \mu, \sigma)$ is the Gaussian density with mean μ and standard deviation σ,

$$g(y; \mu, \sigma) = \frac{1}{\sigma\sqrt{2\pi}} \exp\left\{-\frac{1}{2}\left(\frac{(y-\mu)}{\sigma}\right)^2\right\}, \quad (13.2)$$

and $p_i, \ldots, p_k$ are the mixing probabilities.

The parameters $p_1, \cdots, p_k$, $\mu_1, \cdots, \mu_k$ and $\sigma_1, \cdots, \sigma_k$ are usually estimated by maximum likelihood. Standard errors can be obtained in the usual way from the observed information matrix (i.e., the inverse of the Hessian matrix, the matrix of second derivatives of the log-likelihood). Determining the number of components in the mixture, k, is more problematic. This question is not considered here.

For a short introduction to finite mixture modeling, see Everitt (1996); a more comprehensive account is given in McLachlan and Basford (1988).

Table 13.1 Age of onset of schizophrenia in 99 women. The data are in *female.dat*

Age of onset			
20	30	21	23
30	25	13	19
16	25	20	25
27	43	6	21
15	26	23	21
23	23	34	14
17	18	21	16
35	32	48	53
51	48	29	25
44	23	36	58
28	51	40	43
21	48	17	23
28	44	28	21
31	22	56	60
15	21	30	26
28	23	21	20
43	39	40	26
50	17	17	23
44	30	35	20
41	18	39	27
28	30	34	33
30	29	46	36
58	28	30	28
37	31	29	32
48	49	30	

13.3 Analysis using Stata

Stata has a set of commands, called ml, that can be used to maximize a user-specified log-likelihood using a modified Newton-Raphson algorithm. The algorithm is iterative. Starting with initial parameter estimates, the program evaluates the first and second derivatives of the log-likelihood in order to update the parameters in such a way that the log-likelihood is likely to increase. The updated parameters are then used to re-evalute the derivatives, etc. until the program has converged to the maximum. The user needs to write a program that enables the ml program to evaluate the log-likelihood and its derivatives.

There are four alternative methods available with ml, deriv0, deriv1, deriv2, and lf. The method deriv0 does not require the user's program to evaluate any derivatives of the log-likelihood, deriv1 requires first derivatives only, and deriv2 requires both first and second derivatives. When deriv0 is chosen, first and second derivatives are determined numerically; this makes this alternative slower and less accurate than `deriv1` or `deriv2`.

The simplest approach to use is `lf`, which also does not require any derivatives to be programmed. Instead, the structure of most likelihood problems is used to increase both the speed and accuracy of the numerical estimation of first and second derivatives of the log-likelihood function. Whereas `deriv0` to `deriv2` can be used for any maximum likelihood problem, method `lf` can only be used if the likelihood satisfies the following two criteria:

1. The observations in the dataset are independent, i.e., the log-likelihood is the sum of the log-likelihood functions of the observations.

2. The log-likelihood function for the ith oservation has a *linear form,* i.e., it is a function of linear predictors of the form $\eta_i = x_{1i}\beta_i + \cdots x_{ki}\beta_k$.

The first restriction is usually met, an exception being clustered (e.g., repeated measures) data. The second restriction is not as severe as it appears because there may be several linear predictors, as we shall see later.

We shall eventually use method `lf` to fit a mixture of normals to the age of onset data, but will introduce the `ml` procedure by a series of more simple steps. To begin with, we fit a normal distribution with standard deviation fixed at 1 to a set of simulated data. A (pseudo)-random sample from the normal distribution with mean 5 and variance 1 can be obtained using the instructions

```
set obs 100
set seed 12345678
gen y = invnorm(uniform())+5
```

where the purpose of the `seed` command is simply to ensure that the same data will be generated each time. We use `summarize` to confirm that the sample has a mean close to 5 and a standard deviation close to 1.

```
summarize y
```

Variable	Obs	Mean	Std. Dev.	Min	Max
y	100	5.002311	1.053095	2.112869	7.351898

A set of commands to estimate the mean for a fixed standard deviation of one is given below.

```
eq xb: y

program define mixing0
        local lj "'1'"
        local xb "'2'"

        tempvar f s
        quietly gen double 's' = 1

        quietly gen double 'f' = /*
        */ exp(- .5*(($S_mldepn-'xb')/'s')^2)/
           (sqrt(2*_pi)*'s')
        quietly replace 'lj' = ln('f')
end

ml begin
ml function mixing0
ml method lf
ml model b = xb
ml sample mysamp
ml maximize f V, trace
ml post myest
ml mlout myest
```

Each command is explained below.

eq xb: y defines an equation called xb in which y is the dependent variable
and there are no independent variables besides a constant. The equation
defines the linear predictor to be used in the likelihood as $1 \times \beta$, so that
a single parameter β is to be estimated that represent the coefficient of 1,
the constant. This parameter is of course simply the mean. The name of
the dependent variable is stored in the global macro $S_mldepn while ml is
running.

program define mixing0 defines the program to evaluate the log-likelihood
functions when called by ml. The program has two arguments '1' and '2',
which represent, respectively, the variable that will contain the log-likelihood
functions computed by mixing0, and the variable that contains the linear
predictor computed by ml before mixing0. After the first two commands,
the local macros 'lj' and 'xb' contain the names of these variables. Two
temporary variables, 's' and 'f', are used to hold intermediate results in
the calculation of the log-likelihood functions. For this problem, the log-
likelihood functions are simply the logs of the normal densities evaluated for
each observation. The temporary variable 'f' contains these normal densities (see equation (13.2)), which are a function of the difference between

the dependent variable $S_mldepn and 'xb'. The final command `replace`
'lj' = ln('f') places the log-likelihood functions into the variable 'lj'
as required by the calling program. Note that all variables defined within a
program to be used with `ml` should be of storage type "double" to enable `ml`
to estimate accurate numerical derivatives. Scalars should be used instead of
local macros to hold constants as scalars have higher precision.

The following `ml` instructions are always used for maximum likelihood estimation and should be given in the same order as here.

`ml begin` is required before other `ml` commands can be used.

`ml mixing0` specifies that the program `mixing0` should be used to evaluate
the log-likelihood contributions.

`ml method lf` states the method `lf` should be used to maximize the likelihood.

`ml model b = xb` tells `ml` that the equation to be estimated is `xb` and that
the matrix holding the parameter estimates is to be called **b**.

`ml sample mysamp` causes `ml` to create a variable *mysamp*, which is 1 for all
observation used in the estimation; that is, for the sample of observations
with complete data on all variables used in the model.

`ml maximise f V, trace` causes `ml` to start running and to store the maximized log-likelihood function in the scalar **f** and the covariance matrix
of the estimated parameters in the matrix **V**. The option `trace` causes
parameter estimates to be displayed for each iteration.

`ml post myest` saves the results under the name "myest," in the same form
as Stata's own estimation commands. The results can then be accessed in
the same way as after, for example, `regress`.

`ml mlout myest` displays the estimates and their standard errors etc. in a
table using the same style as Stata's own estimation commands.

Running the above commands gives the results shown in Display 13.1.

Starting from an initial estimate of 0, the estimated mean quickly converges
to the sample mean of 5.002311. The standard error was computed using the
inverse of the second derivative of the log-likelihood function.

We now need to extend the program step by step until it can be used to
estimate a mixture of two Gaussians. The first step is to allow the standard
deviation to be estimated. Since this parameter does not contribute linearly
to the linear predictor used to estimate the mean, we need to define another
linear predictor by specifying another equation with no dependent variable; for
example,

```
eq lsd:
```

This equation is added to the model using

```
   0
Iteration 0:  Log Likelihood = -1397.9454
   5.0023106
Iteration 1:  Log Likelihood = -146.78981
   5.0023106
Iteration 2:  Log Likelihood = -146.78981

                                        Number of obs    =       100
                                        Model chi2(0)    =         .
                                        Prob > chi2      =         .
Log Likelihood =   -146.7898087


-----------------------------------------------------------------------
      y |     Coef.   Std. Err.      z    P>|z|     [95% Conf. Interval]
--------+--------------------------------------------------------------
  _cons |   5.002311         .1   50.023  0.000      4.806314    5.198307
-----------------------------------------------------------------------
```

Display 13.1

```
ml model b = xb lsd, depv(10)
```

where the option depv(10) informs ml that the first equation has one dependent variable and the second equation has none.

We also need to modify the function mixing0 so that it accepts the name of the variable containing the linear predictor for the second equation as the third argument '3' and uses this as the current estimate of the standard deviation.

```
program define mixing1
        local lj "'1'"
        local xb "'2'"
        local ls "'3'"

        tempvar f s
        quietly gen double 's' = exp('ls')

        quietly gen double 'f' = /*
        */ exp(- .5*(($S_mldepn-'xb')/'s')^2)/
           (sqrt(2*_pi)*'s')
        quietly replace 'lj' = ln( 'f')
end
```

Take the exponential of the linear predictor for equation lsd and treat this as the standard deviation in the evaluation of the log-likelihood because this ensures that the standard deviation is positive. The estimated parameter should then be interpreted as the log of the standard deviation.

Making these changes gives the output shown in Display 13.2

The standard deviation estimate is obtained by taking the exponential of the estimated coefficient of _cons in equation lsd. Instead of typing disp

```
  0   0
Iteration 0:  Log Likelihood = -1397.9454
(nonconcave function encountered)
   4.141647   0.17154821
Iteration 1:  Log Likelihood = -174.28158
(unproductive step attempted)
   4.7928262   0.1931215
Iteration 2:  Log Likelihood = -150.00477
   4.9797998   0.06642451
Iteration 3:  Log Likelihood = -146.62524
   5.0021498   0.0470101
Iteration 4:  Log Likelihood =  -146.5647
   5.0023106   0.04670838
Iteration 5:  Log Likelihood = -146.56469
   5.0023105   0.04670833
Iteration 6:  Log Likelihood = -146.56469

                                        Number of obs   =        100
                                        Model chi2(0)   =          .
                                        Prob > chi2     =          .
Log Likelihood =   -146.5646867


-----------------------------------------------------------------------
      y |      Coef.    Std. Err.       z     P>|z|    [95% Conf. Interval]
--------+--------------------------------------------------------------
xb      |
  _cons |   5.002311    .1047816    47.740   0.000     4.796942    5.207679
--------+--------------------------------------------------------------
lsd     |
  _cons |   .0467083    .0707107     0.661   0.509     -.091882    0.1852987
-----------------------------------------------------------------------
```

Display 13.2

exp(0.0467083), we can use the following syntax for accessing coefficients and
their standard errors:

```
disp [lsd]_b[_cons]
```

> .04670833

```
disp [lsd]_se[_cons]
```

> .07071068

When the coefficient is required, the "_b" can be omitted from the expression.
We therefore obtain the required standard deviation using the command

```
disp exp([lsd][_cons])
```

> 1.0478163

This is smaller than the sample standard deviation from `summarize` because the maximum likelihood estimate of the standard deviation is given by

$$\hat{\sigma} = \sqrt{\frac{1}{n} \sum_{i=1}^{n} (y_i - \bar{y})^2} \tag{13.3}$$

where n is the sample size, whereas the factor $\frac{1}{n-1}$ is used in `summarize`. Since n is 100 in this case, the maximum likelihood estimate must be blown up by a factor $\sqrt{100/99}$ to obtain the sample standard deviation:

```
disp exp([lsd][_cons])*sqrt(100/99)
```

```
1.053095
```

The program can now be extended to estimate a mixture of two Gaussians. In order to be able to test it on data from a known distribution, we generate a sample from a mixture of two Gaussians with variances of 1 and means of 0 and 5, respectively, and with mixing probabilities $p_1 = p_2 = 0.5$. This can be done in two stages; first, randomly allocate observations to groups (variable z) with probabilities p_1 and p_2 and then sample from the different component densities according to group membership.

```
set obs 300
set seed 12345678
gen z = cond(uniform()<0.5,1,2)
gen y = invnorm(uniform())
replace y = y + 5 if z==2
```

We now need five equations, one for each parameter to be estimated: μ_1, μ_2, σ_1, σ_2, and p_1 (since $p_2 = 1 - p_1$). As before, we can ensure that the estimated standard deviations are positive by taking the exponential inside the program. The mixing proportion p_1 must lie in the range $0 \leq p_1 \leq 1$. One way of ensuring this is to interpret the linear predictor as representing the log odds (see Chapter 6) so that p is obtained from the linear predictor of the log odds, `lo1` using the transformation $1/(1+\exp(-lo1))$. The program `mixing1` now becomes

```
program define mixing2
        local lj  "`1'"
        local xb1 "`2'"
        local xb2 "`3'"
        local lo1 "`4'"
        local ls1 "`5'"
        local ls2 "`6'"
```

```
tempvar f1 f2 p s1 s2

quietly gen double 'p' = 1/(1+exp(-'lo1'))
quietly gen double 's1' = exp('ls1')
quietly gen double 's2' = exp('ls2')

quietly gen double 'f1' = /*
        */ exp(- .5*(($S_mldepn-'xb1')/'s1')^2)/
           (sqrt(2*_pi)*'s1')
quietly gen double 'f2' = /*
        */ exp(- .5*(($S_mldepn-'xb2')/'s2')^2)/
           (sqrt(2*_pi)*'s2')
quietly replace 'lj' = ln( 'p'*'f1' + (1-'p')*'f2')
end
```

Stata simply uses the value 0 as the initial value for all parameters. When fitting a mixture distribution, it is not advisable to start with the same initial value for both means. Therefore, the initial values should be set. This can be done by defining a matrix with one row which contains the initial values of some (or all) of the parameters. For this example, five equations are defined.

```
eq xb1: y
eq xb2:
eq lsd1:
eq lsd2:
eq lo1:
```

Then define a matrix b0 of initial values as follows:

```
mat b0 = (1,6,0,0.2,-0.2)
```

where we intend the first two values to be initial values for the means, the third to be the initial log odds, and the third and fourth to be the initial values of the logs of the standard deviations. We can specify which parameter is to take on which of the initial values by defining equation labels for the columns of b0 using the command

```
mat coleq b0 = xb1 xb2 lo1 lsd1 lsd2
```

Since there may be more than one parameter per equation, we must also specify to which parameters in each equation the columns of b0 should correspond. This is done by labeling the columns with the variable names of the parameters, in this example the constant, _cons;

```
mat colnames b0 = _cons _cons _cons _cons _cons
```

In the ml model statement, specify that initial estimates are to be taken from this matrix using the option `from(b0)`. The likelihood statements therefore are

```
ml begin ml function mixing2
ml method lf
ml model b = xb1 xb2 lo1 lsd1 lsd2, depv(10000) from(b0)
ml sample mysamp
ml maximize f V, trace
ml post myest
ml mlout myest
```

The results are shown in Display 13.3.

```
     1    6    0   0.2  -0.2
Iteration 0:  Log Likelihood = -746.68918
(nonconcave function encountered)
     .41171582   5.2277895    .19201258    .29298216  -0.00062006
Iteration 1:  Log Likelihood =  -641.7476
(unproductive step attempted)
     .12585212   5.097688     .17662589    .11619987  -.04887887
Iteration 2:  Log Likelihood = -626.41822
    -.03279443   5.019478     .12642401   -.01148219  -.02908023
Iteration 3:  Log Likelihood = -622.33188
    -.04619338   4.9839157    .11204775   -.02788073  -.01776593
Iteration 4:  Log Likelihood = -622.21169
    -.04574516   4.9837169    .11224164   -.027468    -.01734419
Iteration 5:  Log Likelihood = -622.21162
    -.04574518   4.9837169    .11224143   -.027468    -.01734435
Iteration 6:  Log Likelihood = -622.21162
                                    Number of obs    =       300
                                    Model chi2(0)    =
                                    Prob > chi2      =         .
Log Likelihood =    -622.2116180
--------------------------------------------------------------------
       y |     Coef.   Std. Err.      z     P>|z|     [95% Conf. Interval]
---------+----------------------------------------------------------
xb1      |
   _cons |  -.0457452   .0806427   -0.567   0.571    -.203802    .1123117
---------+----------------------------------------------------------
xb2      |
   _cons |   4.983717   .0863492   57.716   0.000    4.814476   5.152958
---------+----------------------------------------------------------
lo1      |
   _cons |   .1122414   .1172032    0.958   0.338   -.1174726   .3419554
---------+----------------------------------------------------------
lsd1     |
   _cons |  -.027468    .0627809   -0.438   0.662   -.1505164   .0955804
---------+----------------------------------------------------------
lsd2     |
   _cons |  -.0173443   .0663055   -0.262   0.794   -.1473008   .1126121
--------------------------------------------------------------------
```

Display 13.3

The standard deviations are therefore estimated as

```
disp exp([lsd1][_cons])
```

.97290582

and

```
disp exp([lsd2][_cons])
```

.9828052

and the probability of group 1 is

```
disp 1/(1 + exp(-[lo1][_cons]))
```

.52803094

Estimates of mixing proportion and means and standard deviations, knowing group membership in the sample, are obtained using

```
table z, contents(freq mean y sd y)
```

z	Freq.	mean(y)	sd(y)
1	160	-.0206726	1.003254
2	140	5.012208	.954237

so that the proportion in group 1 is $160/300 = 0.53$. The estimated parameters agree quite closely with the true parameters and with those estimated by knowing the group membership. We can estimate the standard errors of the standard deviations and of the probability using the delta-method (see Dunn, 1989). According to the delta-method, if $y - f(x)$, then approximately, $se(y) = |f'(x)|se(x)$, where $f'(x)$ is the first derivative of $f(x)$ with respect to x evaluated at the estimated value of x. Thus, lsd = ln(sd), so that, by the delta method,

$$se(sd) = sd * se(lsd) . \qquad (13.4)$$

For the probability, lo−ln(p/(1-p)), so that

$$se(p) = p(1 - p) * se(lo) . \qquad (13.5)$$

The approximate standard errors for the standard deviations are therefore

```
scalar x = exp([lsd1][_cons])
disp [lsd1]_se[_cons]*scalar(x)
```

.06107994

```
scalar x = exp([lsd2][_cons])
disp [lsd2]_se[_cons]*scalar(x)
```

> .06516543

and for the probability

```
scalar x = 1/(1 + exp(-[lo1][_cons]))
disp [lo1]_se[_cons]*scalar(x*(1-x))
```

> .0292087

In the above commands, the function *scalar()* was used to make sure that x is interpreted as the scalar we have defined and not as a variable we may happen to have in the dataset. This was not necessary in the program `mixing2` because there a temporary name was used; temporary names look something like __0009J4 and are defined deliberately to be different to any variable name in the dataset.

We can now apply the same program to the age of onset data. The data can be read in using

```
infile y using onset.dat
gen ly=ln(y)
label variable y ''age of onset of schizophrenia''
```

A useful graphical display of the data is a histogram found from

```
graph y, hist bin(10) xlab gap(3)
```

which is shown in Figure 13.1.

Plausible initial values found from an examination of the histogram can be set as

```
mat b0 = (20,45,0,1.7,2.0)
mat coleq b0 = xb1 xb2 lo1 lsd1 lsd2
mat colnames b0 = _cons _cons _cons _cons _cons
```

The output is given in Display 13.4. The means are 24.8 and 46.4 with standard deviations 6.5 and 7.1, respectively, and the mixing proportions are 0.74 and 0.26, respectively. The standard errors can be estimated as before; see exercises. The histogram strongly suggests that there are two subpopulations. In order to test this more formally, we could also fit a single normal distribution and compare the likelihoods. Wolfe (1971) suggests, on the basis of a limited simulation study, that the difference in minus twice the log-likelihood for a model with c components compared with a model of $c + 1$ components has approximately a χ^2 distribution with $2\nu - 2$ degrees of freedom, where ν is the number of extra parameters in the $c + 1$ component mixture. We save the likelihood of the current model in a local macro

Figure 13.1 *Histogram of age of onset of schizophrenia in women.*

```
local 11 = $S_E_11
```

and fit the single normal model using the program mixing1 as follows:

```
matrix b0=(30,1.9)
matrix coleq b0= xb lsd
matrix colnames b0 = _cons _cons
eq xb: y
eq lsd:

ml begin
ml function mixing1
ml method lf
ml model b = xb lsd, depv(10) from(b0)
ml sample mysamp
ml maximize f V, trace
ml post myest
ml mlout myest
```

with the result shown in Display 13.5.

We can now compare the likelihoods using the method proposed by Wolfe:

```
   20   45   0  1.7    2
Iteration 0:  Log Likelihood = -392.22112
(nonconcave function encountered)
 20.620355  44.742803  .35836093   1.826951  2.3205949
Iteration 1:  Log Likelihood = -382.70539
(nonconcave function encountered)
22.891744  44.005878  .70846293  1.8627884    2.118614
Iteration 2:  Log Likelihood = -375.25359
(unproductive step attempted)
 23.118878  43.990973  .71471956  1.7963554  2.1542801
Iteration 3:  Log Likelihood = -374.70377
 24.314983  46.625607  1.0424959  1.8489852   1.981963
Iteration 4:  Log Likelihood = -373.88184
 24.783361  46.391889  1.0284024  1.8744983  1.9537989
Iteration 5:  Log Likelihood = -373.66971
 24.797798  46.447262  1.0344579  1.8777136  1.9549738
Iteration 6:  Log Likelihood = -373.66896
 24.797714  46.446842  1.0344143  1.8776981  1.9550097
Iteration 7:  Log Likelihood = -373.66896

                                    Number of obs   =       99
                                    Model chi2(0)   =        .
                                    Prob > chi2     =        .
Log Likelihood =   -373.6689563
```

---

y	Coef.	Std. Err.	z	P>|z|	[95% Conf. Interval]	
xb1						
_cons	24.79771	1.133546	21.876	0.000	22.576	27.01942
xb2						
_cons	46.44684	2.740875	16.946	0.000	41.07482	51.81886
lo1						
_cons	1.034414	.3697508	2.798	0.005	.309716	1.759112
lsd1						
_cons	1.877698	.1261093	14.889	0.000	1.630528	2.124868
lsd2						
_cons	1.95501	.2569207	7.609	0.000	1.451454	2.458565

Display 13.4

```
local chi2 = 2*('ll'-$S_E_ll)
disp chiprob(4,'chi2')
```

```
.00063994
```

This test confirms that there appear to be two subpopulations.

```
   30   1.9
Iteration 0:   Log Likelihood = -429.14924
   30.199839   2.3213484
Iteration 1:   Log Likelihood = -385.33116
   30.449083   2.4582184
Iteration 2:   Log Likelihood = -383.39806
   30.47479   2.4537569
Iteration 3:   Log Likelihood = -383.39585
   30.474747   2.4537468
Iteration 4:   Log Likelihood = -383.39585

                                    Number of obs    =      99
                                    Model chi2(0)    =       .
                                    Prob > chi2      =       .

Log Likelihood =   -383.3958458

------------------------------------------------------------------------
       y |    Coef.    Std. Err.      z     P>|z|    [95% Conf. Interval]
---------+--------------------------------------------------------------
xb       |
   _cons |   30.47475   1.169045   26.068   0.000    28.18346    32.76603
---------+--------------------------------------------------------------
lsd      |
   _cons |   2.453747   .0710669   34.527   0.000    2.314458    2.593035
------------------------------------------------------------------------
```

Display 13.5

13.4 Exercises

1. Create a do-file with the commands necessary to fit the mixture model.

2. Add commands to the end of the do-file to calculate the standard deviations and mixing probability and the standard errors of these parameters. What are the standard errors of the estimates for the age of onset data?

3. Simulate values from two normals trying out different values for the various parameters. Do the estimated values tend to lie within two estimated standard errors from the 'true' values?

4. Use program mixing1 to fit a linear regression model to the slimming data from Chapter 5 using only *status* as the explanatory variable. Compare the standard deviation with the root mean square error obtained using the command regress.

5. Use the same program again to fit a linear regression model where the variance is allowed to differ between the groups defined by *status*. Is there any evidence for heteroscedasticity? How do the results compare with those of sdtest?

6. Extend the program mixing2 to fit a mixture of 3 normals and test this on simulated data. (Hint: Use transformations $p1 = 1/d$, $p2 = \exp(lo1)/d$, $p3 = \exp(lo2)/d$ where $d = 1 + \exp(lo1) + \exp(lo2)$) and test it on simulated data.)

Appendix: Answers to Selected Exercises

Chapter 1

2. ```
   cd c:\user
   insheet using test.dat, clear
   ```

4. ```
   label define s 1 male 2 female
   label values sex s
   ```

5. ```
   gen id=_n
   ```

6. ```
   rename v1 time1
   rename v2 time2
   rename v3 time3
   ```

   or

   ```
   for v1-v3\ 1-3,ltype(varlist numeric): rename @1  time@2
   ```

7. ```
   reshape long time, i(id) j(occ)
   ```

8. ```
   egen d=mean(time), by(id)
   replace d=(time-d)^2
   ```

9. ```
   drop if occ==3&id==2
   ```

Chapter 2

1. ```
   table depress, contents(mean weight)
   ```

2. ```
   for iq age weight: table life, contents( mean @ sd @)
   ```

3. ```
   lookup mann
   help signrank
   ```

4. ```
   ranksum weight, by(life)
   ```

5. ```
   gen iq1=iq if life==2
   gen iq2=iq if life==1
   label variable iq1 "no"
   label variable iq2 "yes"
   graph iq1 iq2 age, s(dp) xlabel ylabel jitter(2) ll("IQ")
   ```

6. Save the commands in the "Review" window and edit the file using for example notepad. Add the commands given in the do-file template in Section 1.9 and save the file with the extension *.do*. Run the file by typing the command `do filename`.

Chapter 4

1. `infile bp11 bp12 bp13 bp01 bp02 bp03 using bp.raw`

2. ```
sort drug
graph bp, box by(drug)
sort diet
graph bp, box by(diet)
sort biofeed
graph bp, box by(biofeed)
```

4. ```
sort id
merge id using age.dat
anova bp drug diet biofeed age, continuous(age)
```

Chapter 5

1. `anova resp cond*status status cond, sequential`

2. ```
gen dcond=cond-1
gen dstat=status-1
gen dinter=dcond*dstat
regress resp dcond dstat dinter
```

3. `xi: regress resp i.cond*i.status`

4. ```
char cond[omit] 2
char status[omit] 2
xi: regress resp i.cond*i.status
```

Chapter 6

1. `ologit outc therapy sex [fweight=fr], table`

2. a. `ologit depress life`

   b. `logistic life depress`

3. Even if we use very lenient inclusion and exclusion criteria,

   ```
   sw logistic life depress anxiety iq sex sleep, pr(0.2) pe(0.1)
   forward
   ```

   only *depress* is selected. If *depress* is excluded from the list of candidate variables, *anxiety* and *sleep* are selected.

4. ```
bprobit pres tot ck
predict predp
graph predp prop ck, c(l.)
```

## Chapter 7

1. ```
xi: glm resp i.cond i.status, fam(gauss) link(id)
local dev1=$S_E_dev
xi: glm resp i.cond, fam(gauss) link(id)
local dev2=$S_E_dev
local ddev='dev2'-'dev1'
/* F-test equivalent to anova cond status, sequential */
local f=('ddev'/1)/('dev1'/31)
disp 'f'
disp fprob(1,31,'f')
```

2. ```
reg resp status, robust
ttest resp, by(status) unequal
```

3. ```
gen cleth=class*ethnic
glm days class ethnic cleth, fam(poiss) link(log)
```

4. ```
glm days class ethnic if stres<4, fam(poiss) link(log)
```

   or, assuming the data has not been sorted,

   ```
glm days class ethnic if n~=72, fam(poiss) link(log)
```

5. Try lookup to see if a suitable program has been included in an STB, visit the Stata Web site to see if one is included as a "cool ado" or visit the archive of ado files sent to Statalist, at
   http://ideas.uqam.ca/ideas/data/bocbocode.html. Copy the ado and help file (extension .hlp) to your personal ado directory (usually c:\ado, but you can use the command adopath to list directories where Stata searches for ado-files) and simply use the ado-file according to the instructions given in the help file.

6. ```
gen abs=cond(days>=14,1,0)
glm abs class ethnic, fam(binomial) link(logit)
glm abs class ethnic, fam(binomial) link(probit)
```

7. ```
logit abs class ethnic, robust
probit abs class ethnic, robust
bs "logit abs class ethnic" "_b[class] _b[ethnic]", reps(500)
bs "probit abs class ethnic" "_b[class] _b[ethnic]", reps(500)
```

## Chapter 8

1. ```
graph dep1-dep6, box by(group)
```

2. We can obtain the mean over visits for subjects with complete data using
the simple command

```
gen av2 = (dep1+dep2+dep3+dep4+dep6)/6
```

The t-tests are obtained using

```
ttest av2, by(group)
ttest av2, by(group) unequal
```

3.
```
egen sd = rsd(dep1-dep6)
gen stdav= avg/sd
ttest stdav, by(group)
```

4. a.
```
gen diff =  avg-pre
   ttest diff, by(group)
```

   b.
```
anova avg group pre, continuous(pre)
```

Chapter 9

2. a.
```
reg dep group pre visit, robust cluster(subj)
```
   b.
```
bs "reg dep group pre visit" "_b[group] _b[pre] _b[visit]", /*
      */ cluster(subj) reps(500)
```

5.
```
expand 2 if week==1
sort subj week
qui by subj: replace week=0 if _n==1
replace y=baseline/4 if week==0
gen post=cond(week==0,0,1)
xi: xtgee y i.treat*i.post age , i(subj) t(week) corr(exc) /*
          */ family(pois) scale(x2) eform
```

Chapter 10

1.
```
infile v1-v2 using estrogen.dat, clear
gen str8 _varname="ncases1" in 1
replace _varname="ncases0" in 2
xpose,clear
gen conestr=2-_n
reshape long ncases, i(conestr) j(casestr)
expand ncases
sort casestr conestr
gen caseid=_n
expand 2
sort caseid
quietly by caseid: gen control=_n-1 /*control=1, (0=case) */
gen estr=0
```

```
replace estr=1 if control==0&casestr==1
replace estr=1 if control==1&conestr>0
gen cancer=0
replace cancer=1 if control==0
clogit cancer estr, group(caseid) or
```

2. ```
table exposed, contents(sum num sum py)
iri 17 28 2768.9 1857.5
```

3. ```
xi: poisson num i.age*exposed, e(py) irr
testparm IaX*
```

The interaction is not statistically significant at the 5% level.

4. ```
infile subj y1 y2 y3 y4 treat baseline age using chemo.dat
reshape long y, i(subj) j(week)
expand 2 if week==1
sort subj week
qui by subj: replace week-0 if _n--1
replace y=baseline if week==0
gen post=cond(week==0,0,1)
gen ltime=log(cond(week==0,8,2))
xi: xtgee y i.treat*i.post age , i(subj) t(week) corr(exc) /*
          */ family(pois) scale(x2) offset(ltime)
```

## Chapter 11

1. We consider anyone still at risk after 450 days as being censored at 450 days and therefore need to make the appropriate changes to *status* and *time* before running Cox regression.

```
replace status=0 if time>450
replace time=450 if time>450
stset time status
stcox dose prison clinic
```

2. The model is fitted using

```
gen dose_cat=0 if dose~=.
replace dose_cat=1 if dose>=60
replace dose_cat=2 if dose>=80
xi: stcox i.dose_cat i.prison i.clinic, bases(s)
```

The survival curves for no prison record, clinic 1 are obtained using

```
gen s0 = s if dose_cat==0
gen s1 = s^(exp(_b[Idose__1]))
gen s2 = s^(exp(_b[Idose__2]))
graph s0 s1 s2 time, sort c(JJJ) s(...) xlab ylab gap(3)
```
```

Note that the baseline survival curve is the survival curve for someone whose covariates are all zero. If we had used `clinic` instead of `i.clinic` above, this would have been meaningless; we would have had to exponentiate $s0$, $s1$ and $s2$ by `_b[clinic]` to calculate the survival curves for clinic 1.

3. Treating dose as continuous:

```
gen clindose=clinic*dose
stcox dose prison clinic clindose
```

Treating dose as categorical:

```
xi: stcox i.dose_cat*i.clinic i.prison
testparm IdX*
```

5.
```
gen tpris=prison*(t-504)/365.25
stcox dose prison tpris, strata(clinic)
lrtest, saving(0)
stcox dose prison, strata(clinic)
lrtest
```

6. Replace the section defining *num* by:

```
capture program drop rednum
program define rednum
      local nt1=rowsof(t1)
      local i=1
      while 'i'<='nt1'{
            qui replace num=num-1 /*
               */ if clinic==1&time<=t1['nt1'-'i'+1,1]
               local i='i'+1
      }
      local nt2=rowsof(t2)
      local i=1
      while 'i'<='nt2'{
            qui replace num=num-1 /*
               */ if clinic==2&time<=t2['nt2'-'i'+1,1]
               local i='i'+1
      }
end
local nt1=rowsof(t1)
local nt2=rowsof(t2)
gen num=cond(clinic==1,'nt1','nt2')
rednum
replace num=num+1
```

Chapter 12

1. `factor l500-r4000, pc cov`

3.
```
capture drop pc*
factor l500-r4000, pc
score pc1-pc5
graph pc1-pc5, matrix
```

5.
```
infile str10 town so2 temp manuf pop wind precip days
using usair.dat
factor temp manuf pop wind precip days,pc
score pc1 pc2
graph pc2 pc1, symbol([town]) ps(150) gap(3) xlab ylab
```

6.
```
regress so2 pc1 pc2
gen line=pc1*_b[pc2]/_b[pc1]
graph pc2 line pc1, s([town]i) c(.l) ps(150) gap(3) xlab ylab
```

Chapter 13

2.
```
scalar x = exp([lsd1][_cons])
disp "st. dev. 1 = " scalar(x) ", standard error = " /*
    */ [lsd1]_se[_cons]*scalar(x)

scalar x = exp([lsd2][_cons])
disp "st. dev. 2 = " scalar(x) ", standard error = " /*
    */ [lsd2]_se[_cons]*scalar(x)

scalar x = 1/(1 + exp(-[lo1][_cons]))
disp "probability = " scalar(x) ", standard error = "/*
    */ [lo1]_se[_cons]*scalar(x*(1-x))
```
giving the results
```
st. dev. 1 = 6.5384366, standard error = .82455744
st. dev. 2 = 7.0639877, standard error = 1.8148844
probability = .7377708, standard error = .07153385
```

4. The only thing that is different from fitting a normal distribution with constant mean is that the mean is now a linear function of *status* so that eq xb changes as shown below.
```
infile cond status resp using slim.dat, clear

matrix b0=(0)
eq xb: res status
eq lsd:
ml begin
ml function mixing1
ml method lf
ml model b = xb lsd, depv(10) from(b0)
ml sample mysamp
```

```
ml maximize f V, trace
ml post myest
ml mlout myest
```

The mean square error of the regression

```
regress resp status
```

is equal to the sum of squares divided by the degrees of freedom, $n-2$. The maximum likelihood estimate is equal to the sum of squares divided by n. We can therefore get the root mean square error of the regression result using

```
disp exp([lsd][_cons])*sqrt(_N/(_N-2))
```

Note that the standard error of the regression coefficients needs to be corrected by the same factor, i.e.,

```
disp [xb]_se[status]*sqrt(_N/(_N-2))
```

5. Repeat the procedure above but replace the equation for the standard deviation by

```
eq lsd: status
```

The effect of *status* on the standard deviation is significant $(p = 0.003)$ which is not too different from the result of

```
sdtest resp, by(status)
```

6.
```
eq xb1: y
eq xb2:
eq xb3:
eq lsd1:
eq lsd2:
eq lsd3:
eq lo1:
eq lo2:

program define mixing3
        local lj  "'1'"
        local xb1 "'2'"
        local xb2 "'3'"
        local xb3 "'4'"
        local lo1 "'5'"
        local lo2 "'6'"
        local ls1 "'7'"
        local ls2 "'8'"
        local ls3 "'9'"
```

```
        tempvar f1 f2 f3 p1 p2 p3 s1 s2 s3 d

quietly{
        gen double 'p1' = 1.0
        gen double 'p2' = exp('lo1')
        gen double 'p3' = exp('lo2')
        gen double 'd' = 'p1' + 'p2' + 'p3'
        replace 'p1' = 'p1'/'d'
        replace 'p2' = 'p2'/'d'
        replace 'p3' = 'p3'/'d'

        gen double 's1' = exp('ls1')
        gen double 's2' = exp('ls2')
        gen double 's3' = exp('ls3')

        gen double 'f1' = /*
            */ exp(- .5*(($S_mldepn-'xb1')/'s1')^2)/
               (sqrt(2*_pi)*'s1')
        gen double 'f2' = /*
            */ exp(- .5*(($S_mldepn-'xb2')/'s2')^2)/
               (sqrt(2*_pi)*'s2')
        gen double 'f3' = /*
            */ exp(- .5*(($S_mldepn-'xb3')/'s3')^2)/
               (sqrt(2*_pi)*'s3')

        replace 'lj' = ln( 'p1'*'f1' + 'p2'*'f2' +
                           'p3'*'f3')
}
end

ml begin
ml function mixing3
ml method lf
ml model b = xb1 xb2 xb3 lo1 lo2 lsd1 lsd2 lsd3, /*
            */ depv(10000000) from(b0)
ml sample mysamp
ml maximize f V, trace
ml post myest
ml mlout myest
```

References

M. Aitkin. The analysis of unbalanced cross-classifications (with discussion). *Journal of the Royal Statistical Society, A,* 41:195–223, 1978.

D. A. Binder. On the variances of asymptotically normal estimators from complex surveys. *International Statistical Review,* 51:279–292, 1983.

D. R. Boniface. *Experimental Design and Statistical Methods.* Chapman & Hall, London, 1995.

J. Caplehorn and J. Bell. Methadone dosage and the retention of patients in maintenance treatment. *The Medical Journal of Australia,* 154:195–199, 1991.

S. Chatterjee and B. Price. *Regression Analysis by Example (2nd edition).* Wiley, New York, 1991.

D. Clayton and M. Hills. *Statistical Models in Epidemiology.* Oxford University Press, Oxford, 1993.

D. Collett. *Modelling Binary Data.* Chapman & Hall, London, 1991.

D. Collett. *Modelling Survival Data in Medical Research.* Chapman & Hall, London, 1994.

R. D. Cook. Detection of influential observations in linear regression. *Technometrics,* 19:15–18, 1977.

R. D. Cook. Influential observations in linear regression. *Journal of the American Statistical Association,* 74:164–174, 1979.

R. D. Cook and S. Weisberg. *Residuals and Influence in Regression.* Chapman & Hall, London, 1982.

P. J. Diggle, K.-Y. Liang, and S. L. Zeger. *Analysis of Longitudinal Data.* Oxford University Press, Oxford, 1994.

G. Dunn. *Design and Analysis of Reliability Studies.* Oxford University Press, Oxford, 1989.

B. Efron and R. Tibshirani. *An Introduction to the Bootstrap.* Chapman & Hall, London, 1993.

B. S. Everitt. *Cluster Analysis.* E. Arnold, London, 1993.

B. S. Everitt. *Statistical Methods for Medical Investigations.* E. Arnold, London, 1994.

B. S. Everitt. The analysis of repeated measures: a practical review with examples. *The Statistician,* 44:113–135, 1995.

B. S. Everitt. An introduction to finite mixture distributions. *Statistical Methods in Medical Research,* 5:107–127, 1996.

B. S. Everitt and G. Der. *A Handbook for Statistical Analyses Using SAS.* Chapman & Hall, London, 1996.

B. S. Everitt and G. Dunn. *Applied Multivariate Analysis.* E. Arnold, London, 1991.

B. S. Everitt and S. Rabe-Hesketh. *The Analysis of Proximity Data.* E. Arnold, London, 1997.

H. Goldstein. *Multilevel Statistical Models.* E. Arnold, London, 1995.

W. Gould. Estimating a Cox model with continuously time-varying parameter. In Stata FAQs, College Station, TX: Stata Corporation, 1997.

A. J. P. Gregoire, R. Kumar, B. S. Everitt, A. F. Henderson, and J. W. W. Studd. Transdermal oestrogen for the treatment of severe post-natal depression. *The Lancet,* 347:930–934, 1996.

L. C. Hamilton. *Statistics with Stata 5.* Duxbury Press, Belmont, CA, 1998.

D. J. Hand and M. Crowder. *Practical Longitudinal Data Analysis.* Chapman & Hall, London, 1996.

D. J. Hand, F. Daly, A. D. Lunn, K. J. McConway, and E. Ostrowski. *A Handbook of Small Data Sets.* Chapman & Hall, London, 1994.

W. Holtbrugge and M. Schumacher. Analysis of complex statistical variables into principal components. *Journal of Education Psychology,* 24:417–441, 1933.

W. Holtbrugge and M. Schumacher. A comparison of regression models for the analysis of ordered categorical data. *Applied Statistics,* 40:249–259, 1991.

J. E. Jackson. *A User's Guide to Principal Components.* Wiley, New York, 1991.

R. I. Jennrich and M. D. Schluchter. Unbalanced repeated measures models with unstructured covariance matrices. *Biometrics,* 42:805–820, 1986.

P. Lachenbruch and R. M. Mickey. Estimation of error rates in discriminant analysis. *Technometrics,* 10:1–11, 1986.

R. R. J. Lewine. Sex differences in schizophrenia: timing or subtypes? *Psychological Bulletin,* 90:432–444, 1981.

K.-Y. Liang and S. L. Zeger. Longitudinal data analysis using generalised linear models. *Biometrika,* 73:13–22, 1986.

C. L. Mallows. Some comments on C_p. *Technometrics,* 15:661–675, 1986.

B. F. J. Manley. *Randomisation, Bootstrap and Monte Carlo Methods in Biology.* Chapman & Hall, London, 1997.

J. N. S. Mathews, D. G. Altman, M. J. Campbell, and P. Royston. Analysis of serial measurements in medical research. *British Medical Journal,* 300:230–235, 1990.

S. E. Maxwell and H. D. Delaney. *Designing Experiments and Analysing Data.* Wadsworth, California, 1990.

P. McCullagh and J. A. Nelder. *Generalized Linear Models.* Chapman & Hall, London, 1989.

R. J. McKay and N. A. Campbell. Variable selection techniques in discriminant analysis. I description. *British Journal of Mathematical and Statistical Psychology,* 35:1–29, 1982.

R. J. McKay and N. A. Campbell. Variable selection techniques in discriminant analysis. II allocation. *British Journal of Mathematical and Statistical Psychology,* 35:30–41, 1982.

G. J. McLachlan and K. E. Basford. *Mixture Models; Inference and Application to Clustering.* Marcel Dekker, New York, 1988.

J. A. Nelder. A reformulation of linear models. *Journal of the Royal Statistical Society, A,* 140:48–63, 1977.

K. Pearson. On lines and planes of closest fit to points in space. *Philosophical Magazine,* 2:559–572, 1901.

J. O. Rawlings. *Applied Regression Analysis.* Wadsworth Books, California, 1988.

K. J. Rothman. *Modern Epidemiology.* Little, Brown & Company, Boston, 1986.

D. L. Sackett, R. B. Haynes, G. H. Guyatt, and P. Tugwell . *Clinical Epidemiology.* Little Brown & Company, Massachusetts, 1991.

R. R. Sokal and F. J. Rohlf. *Biometry.* W. H. Freeman, San Francisco, 1981.

P. Sprent. *Applied Nonparametric Statistical Methods.* P. Sprent, Chapman & Hall, London, 1993.

B. Sribney. What are the advantages of using robust variance estimators over standard maximum-likelihood estimators in logistic regression. In Stata FAQs, College Station, TX: Stata Corporation, 1998.

Getting Started with Stata. Stata Press, College Station, TX, 1997.

Stata Reference Manual. Stata Press, College Station, TX, 1997.

Stata User's Guide. Stata Press, College Station, TX, 1997.

P. F. Thall and S. C. Vail. Some covariance models for longitudinal count data with overdispersion. *Biometrics,* 46:657–671, 1990.

R. W. M. Wedderburn. Quasilikelihood functions, generalised linear models and the Gauss-Newton method. *Biometrika,* 61:439–47, 1974.

B. J. Winer. *Statistical Principles in Experimental Design.* McGraw-Hill, New York, 1971.

J. H. Wolfe. *A Monte Carlo study of the sampling distribution of the likelihood ratio for mixtures of multinormal distributions,* volume STB72-2 of *Technical Bulletin.* Naval Personnel and Training Research Laboratory, San Diego, 1971.

Index